IOL POWER

IOL POWER

Kenneth J. Hoffer, MD
Clinical Professor of Ophthalmology, Jules Stein Eye Institute
University of California, Los Angeles
Founding President, American Society of Cataract and Refractive Surgery
Founding Editor, Journal of Cataract & Refractive Surgery
Past-President, IOL Power Club (2005-2007)
St. Mary's Eye Center
Santa Monica, CA

www.slackbooks.com

ISBN: 978-1-55642-988-0

Copyright © 2011 by SLACK Incorporated

All rights reserved. No part of this book may be reproduced, stored in a retrieval system or transmitted in any form or by any means, electronic, mechanical, photocopying, recording or otherwise, without written permission from the publisher, except for brief quotations embodied in critical articles and reviews.

The procedures and practices described in this book should be implemented in a manner consistent with the professional standards set for the circumstances that apply in each specific situation. Every effort has been made to confirm the accuracy of the information presented and to correctly relate generally accepted practices. The authors, editor, and publisher cannot accept responsibility for errors or exclusions or for the outcome of the material presented herein. There is no expressed or implied warranty of this book or information imparted by it. Care has been taken to ensure that drug selection and dosages are in accordance with currently accepted/recommended practice. Due to continuing research, changes in government policy and regulations, and various effects of drug reactions and interactions, it is recommended that the reader carefully review all materials and literature provided for each drug, especially those that are new or not frequently used. Any review or mention of specific companies or products is not intended as an endorsement by the author or publisher.

SLACK Incorporated uses a review process to evaluate submitted material. Prior to publication, educators or clinicians provide important feedback on the content that we publish. We welcome feedback on this work.

Published by: SLACK Incorporated
6900 Grove Road
Thorofare, NJ 08086 USA
Telephone: 856-848-1000
Fax: 856-848-6091
www.slackbooks.com

Contact SLACK Incorporated for more information about other books in this field or about the availability of our books from distributors outside the United States.

Hoffer, Kenneth J.
  IOL power / Kenneth J. Hoffer.
      p. ; cm.
  Intraocular lens power
  Includes bibliographical references and index.
  ISBN 978-1-55642-988-0 (alk. paper)
 1. Intraocular lenses. I. Title. II. Title: Intraocular lens power.
  [DNLM: 1. Lenses, Intraocular. 2. Biometry. WW 358]
  RE988.H64 2011
  617.7'524--dc22
                                    2010053713

For permission to reprint material in another publication, contact SLACK Incorporated. Authorization to photocopy items for internal, personal, or academic use is granted by SLACK Incorporated provided that the appropriate fee is paid directly to Copyright Clearance Center. Prior to photocopying items, please contact the Copyright Clearance Center at 222 Rosewood Drive, Danvers, MA 01923 USA; phone: 978-750-8400; web site: www.copyright.com; email: info@copyright.com

Printed in the United States of America.

Last digit is print number: 10 9 8 7 6 5 4 3 2 1

Dedication

I dedicate this book to my wife Marcia (Figure D-1), who has stood beside me for these 40 years while all this cataract/IOL and IOL power history took place.

I also dedicate it to my precious grandchildren: Kaylin (Figure D-2), Anabella (Figure D-3), Wesley (Figure D-4), and Erik Hoffer (Figure D-5). May they forgive me for the time I took from being with them to work on this book.

Kenneth J. Hoffer, MD

Figure D-1. Marcia.

Figure D-2. Kaylin.

Figure D-3. Anabella.

Figure D-4. Wesley.

Figure D-5. Erik.

Contents

Dedication .... *v*
Acknowledgments .... *xi*
About the Author .... *xii*
Contributing Authors .... *xv*
Foreword by Emmanuel Rosen, BSc, MD, FRCSEd, FRCOphth, FRPS .... *xix*
Foreword by Bradley R. Straatsma, MD, JD .... *xxi*

SECTION I   **BASICS AND ACCURATE BIOMETRY** 1

Chapter 1   Section I: Introduction 3
Kenneth J. Hoffer, MD

Chapter 2   Axial Length: Ultrasound: Background 9
Kenneth J. Hoffer, MD

Chapter 3   Immersion Versus Applanation Ultrasound 13
Kenneth J. Hoffer, MD

Chapter 4   Ultrasound Velocities for Axial Length Measurement 17
Kenneth J. Hoffer, MD

Chapter 5   A-Scan Biometry: Immersion Technique 25
H. John Shammas, MD

Chapter 6   A-Scan Biometry: Avoiding Pitfalls With Immersion Technique 29
Karl Ossoinig, MD

Chapter 7   Available A-Scan Instrumentation 39
Kenneth J. Hoffer, MD

Chapter 8   Immersion Using the Prager Shell 53
Thomas C. Prager, PhD, MPH

Chapter 9   Axial Length: Laser Interferometry: Basics of the IOLMaster 63
Wolfgang Haigis, MS, PhD

Chapter 10   IOLMaster Examination 67
Wolfgang Haigis, MS, PhD

Chapter 11   IOLMaster in Difficult Eyes 71
Wolfgang Haigis, MS, PhD

Chapter 12   Axial Length: Laser Interferometry: The LENSTAR LS900
Instrument 75
Kenneth J. Hoffer, MD; H. John Shammas, MD; and Jaime Aramberri, MD

Chapter 13   Corneal Power: Diopters Versus Radius 89
Wolfgang Haigis, MS, PhD

Chapter 14   Corneal Power: Manual Keratometry and Instrumentation 93
Kenneth J. Hoffer, MD

Chapter 15   Automated Keratometry for IOL Power Calculation 97
Jaime Aramberri, MD

Chapter 16 Corneal Power: Corneal Topography for IOL Power Calculation 107
Jaime Aramberri, MD

Chapter 17 Corneal Power: Measuring Corneal Power With the Pentacam 115
Giacomo Savini, MD

Chapter 18 IOL Position: ACD and ELP 119
Kenneth J. Hoffer, MD

Chapter 19 IOL Position: Measuring the ACD by Optical Pachymetry 123
Kenneth J. Hoffer, MD

Chapter 20 IOL Position: Double-AL Method for Scleral Buckle Eyes 131
Kenneth J. Hoffer, MD

Section II FORMULAS AND SPECIAL CIRCUMSTANCES 133

Chapter 21 Section II: Introduction 135
Kenneth J. Hoffer, MD

Chapter 22 Formulas and Programs: Formula History and Basics 137
Kenneth J. Hoffer, MD

Chapter 23 Formulas and Programs: Regression and Theoretic Formulas 143
Kenneth J. Hoffer, MD

Chapter 24 Formulas and Programs: Olsen Formula 149
Thomas Olsen, MD

Chapter 25 Formulas and Programs: Accessing Modern IOL Power Formulas ... 155
Kenneth J. Hoffer, MD

Chapter 26 Formulas and Programs: Formula Personalization 163
Kenneth J. Hoffer, MD

Chapter 27 Special Circumstances: AL Measurement in Staphyloma Eyes 167
H. John Shammas, MD

Chapter 28 Special Circumstances: Silicone Oil–Filled Eyes 169
Wolfgang Haigis, MS, PhD

Chapter 29 Special Circumstances: Unilateral High Myopes and Hyperopes 171
Kenneth J. Hoffer, MD

Chapter 30 Special Circumstances: Penetrating Keratoplasty and
Scarred Corneas 173
Kenneth J. Hoffer, MD

Chapter 31 Special Circumstances: Radial Keratotomy Eyes 175
Giacomo Savini, MD

Chapter 32 Special Circumstances: Post-Laser Refractive Surgery Eyes 179
Kenneth J. Hoffer, MD

Chapter 33 Special Circumstances: Haigis-L IOL Formula 195
Wolfgang Haigis, MS, PhD

Chapter 34 Special Circumstances: Double-K Method 199
           Jaime Aramberri, MD

Chapter 35 Special Circumstances: Influence of Spherical Aberration
           on IOL Power 207
           Sverker Norrby, PhD

Chapter 36 Special Circumstances: Multifocals and Toric IOLs 211
           John Moran, MD, PhD

Chapter 37 Special Circumstances: Pediatric Eyes 215
           Scott K. McClatchey, CAPT, MC, USN, MD

Chapter 38 Special Circumstances: Piggyback IOLs 219
           Kenneth J. Hoffer, MD

Chapter 39 Special Circumstances: Silicone Oil Power 221
           Kenneth J. Hoffer, MD

Chapter 40 Special Circumstances: Effect of IOL Tilt on Astigmatism 223
           Susana Marcos, PhD, FOSA, FEOS

Chapter 41 Special Circumstances: Aniseikonia and Anisometropia 231
           Kenneth J. Hoffer, MD

Chapter 42 Preventing IOL Power Errors 233
           Kenneth J. Hoffer, MD

Chapter 43 Diagnosing and Treating IOL Power Errors 237
           Kenneth J. Hoffer, MD

Chapter 44 Future Directions in IOL Power Calculation:
           Intraoperative Refractive Biometry 241
           Tsontcho [Sean] Ianchulev, MD, MPH

Appendix A: IOL Power Club 247
Financial Disclosures 249
Index 251

Acknowledgments

I first wish to acknowledge the pioneering work in IOL power calculation done by Rob Van der Heidje, PhD (Fig. A-1) from Amsterdam. His original formula, nomogram (Fig. A-2), and disk calculation device (Fig. A-3) (Biometer, Medical Workshop, Groningen, Netherlands) was the earliest to attempt to make it easier for surgeons to select the proper lens power.

Figure A-1. Rob G. L. Van der Heidje, PhD

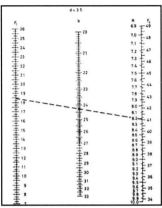

Figure A-2. 1975 Van der Heidje Nomogram using 3.5 mm as a standard ELP.

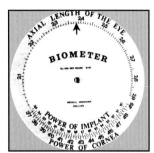

Figure A-3. 1975 Van der Heidje Biometer using 3.5 mm as a standard ELP.

I must personally acknowledge the many people who have helped me over these 37 years, but it is impossible to note them all. I first note the original encouragement of my early (1972-1977) practice partners, John E. Gilmore, MD (deceased) and Donald E. Dickerson, MD; as well as the tireless effort and devotion of my nurses, Patricia Thirsk, RN and Florence Braun, RN in helping to assure accuracy and perfection in the care of my patients and data collection in my research. It would have been difficult in those early years without the support of the UCLA Santa Monica Hospital in giving me quarters for the Eye Lab and providing staff, equipment, and materials to start IOL power calculations the right way. In this endeavor, I have to thank my first A-scan technician, Mr. Don Allen (deceased), who proved technical personnel could do as good a job as a physician. I also thank my many other technicians through the years who were originally trained by Mr. Allen.

On a personal note, I wish to thank John Shammas, MD and Wolfgang Haigis, PhD for their many years of friendship and support; as well as Jack Holladay, MD, a young man in the 70s who helped me prove that theoretic formulas were better than regression and then for keeping me on my toes.

Finally I acknowledge the patience and acceptance of my wife Marcia and our children, Kevin, Jeffrey, and Kristin, who sacrificed time with me while I spent much personal time on these endeavors.

About the Author

Kenneth J. Hoffer, MD was born in New York City; grew up in upstate NY; and received his education at Siena College, Loudonville, NY, SUNY Upstate Medical University in Syracuse, NY, UCLA Santa Monica Hospital, and Wayne State University Kresge Eye Institute in Detroit, MI. He immediately set up practice in Santa Monica, CA in July 1972.

His interest in cataract surgery began early in his clinical career by beginning phacoemulsification after his first two months of practice, and IOL implantation two years later in 1974. Being in an extremely competitive environment, and to create superior patient outcomes, it stimulated him to introduce ultrasound axial length IOL power calculation. He created the American Intra-Ocular Implant Society (now ASCRS) and founded the *Journal of Cataract & Refractive Surgery* (previously AIOISJ). He wrote his original Hoffer formula in 1974 and, based on a pachymetry study in 1982, proposed the first prediction of ELP based on the eye's axial length. The Hoffer Q formula followed in 1993, using a tangent of K as an additional ELP predictor. To ease the calculation of formulas he first used early programmable calculators (HP-65), and in 1993 produced the first program for IOL power calculation for personal computers (Hoffer Programs®). He also participated in the founding of the IOL Power Club in 2005 that has helped stimulate his interest in this subject.

After many years of teaching courses on the subject of IOL power at the Santa Monica Hospital IOL courses (2800 surgeons attended over 10 years); at major meetings such as AAO, ASCRS, and ESCRS; performing research in this field; and publishing papers and book chapters, he was finally stimulated to write his first book on the subject to which he has devoted his professional career. This book is based primarily on the courses he has taught with the additional help of many friends in the field.

Contributing Authors

Jaime Aramberri, MD
BEGITEK Clínica Oftalmológica
OKULAR Clínica Oftalmológica
San Sebastián, Spain
Vitoria-Gasteiz, Spain

Wolfgang Haigis, MS, PhD
Professor of Ophthalmic Biometry
University of Würzburg
Dept. of Ophthalmology
Laboratory for Biometry
Würzburg, Germany

Kenneth J. Hoffer, MD
Clinical Professor of Ophthalmology, Jules Stein Eye Institute
University of California, Los Angeles
Founding President, American Society of Cataract and Refractive
 Surgery
Founding Editor, Journal of Cataract & Refractive Surgery
Past-President, IOL Power Club (2005-2007)
St. Mary's Eye Center
Santa Monica, CA

Tsontcho [Sean] Ianchulev, MD, MPH
Clinical Assistant Professor
University of California, San Francisco
San Francisco, California

Susana Marcos, PhD, FOSA, FEOS
Professor of Research
Instituto de Óptica "Daza de Valdés"
Madrid, Spain

Contributing Authors

Scott K. McClatchey, CAPT, MC, USN, MD
Adjunct Associate Professor of Ophthalmology, Loma Linda
  University
Assistant Professor, Uniformed University of the Health Sciences
Navy Medical Center
San Diego, CA

John R. Moran, PhD, MD
President/CEO
Moran Research and Consulting, Inc.
Houston, TX

Sverker Norrby, PhD
Vice-President, IOL Power Club (2009-2011)
Netherlands

Thomas Olsen, MD
Assistant Professor, dr. med.
University Eye Clinic
Århus Hospital, Denmark
President, Scandinavian Society of Cataract and Refractive Surgery
  (SSCRS)
Board Member, European Society of Cataract & Refractive Surgery
  (ESCRS)
Secretary, IOL Power Club (2005-2011)
Denmark

Karl Ossoinig, MD
Honorary President of SIDUO
Professor Emeritus
Department of Ophthalmology
University of Iowa
Iowa City, IA

Thomas C. Prager, PhD, MPH
Vale Asche Russell Professor of Ophthalmology
University of Texas Medical School
Houston, TX
Clinical Professor, Department of Ophthalmology and Visual Science
University of Texas Medical Branch, Galveston
Clinical Professor, Department of Ophthalmology
Weil Cornell School of Medicine
Houston, TX

Giacomo Savini, MD
Studio Oculistico d'Azeglio
Member, IOL Power Club
Bologna, Italy

H. John Shammas, MD
Clinical Professor of Ophthalmology
The Keck School of Medicine at USC
Los Angeles, CA
Medical Direcor
Shammas Eye Medical Center
Lynwood, CA

Foreword

Ken Hoffer is pre-eminent in the field of ocular biometry. After hundreds of peer-reviewed presentations and contributions to the ophthalmic literature, over 50 chapters in other textbooks, he has at last produced his own authored textbook, *IOL Power,* for what I am sure will be enlightening for all ophthalmologists engaged in lens and refractive surgery.

His academic affiliations attest to his peer recognition and his pleasure in sharing his knowledge as a gifted teacher. He has represented the Department of Ophthalmology and the Jules Stein Eye Institute, and Clinical Faculty, at the University of California in Los Angeles as a Clinical Instructor from 1976 to 1979, Assistant Clinical Professor from 1979 to 1983, Associate Clinical Professor from 1983 to 1990, and Clinical Professor from 1990 to the present time.

His contributions to ophthalmology have been more wide-ranging than his very well-known and heavily utilized Hoffer IOL Power formulae, as he has been a constant innovator in aspects of cataract surgery instrumentation and refractive surgery techniques.

Ken Hoffer was the founding president of ASCRS from 1974 to 1975 and founding editor of the *Journal of Cataract & Refractive Surgery* in 1975. I have particular reason to enjoy Ken's continuing contributions to *JCRS* as author and peer reviewer where he never refuses a request to review and always produces thorough and expert reviews for guidance of the editors.

Since Ken introduced his "Intraocular Lens Power Calculation" lecture at the first Intraocular Lens Symposium held in the USA at Long Beach Memorial Hospital in Long Beach, CA on November 16, 1974, he has been prolific in designing and publishing his formulae and advice for the betterment of ophthalmic patient care. In so doing, he has attracted the thanks of countless ophthalmologists worldwide.

I wish Ken the success that his endeavor, genius, and productivity deserve with success of this volume.

Professor Emanuel Rosen, BSc, MD, FRCSEd, FRCOphth, FRPS
Director, Rosen Eye Associates
Co-Editor, Journal of Cataract and Refractive Surgery
Past-President, European Society of Cataract & Refractive Surgeons (ESCRS)
Manchester, UK

Foreword

Kenneth J. Hoffer, MD, and a group of internationally recognized experts, bring together the physical sciences, mathematical formulas, and clinical experience required to calculate the intraocular lens power for cataract/intraocular lens surgery and refractive phakic/intraocular lens surgery. As a service to readers, the authors provide historical perspective, hone the message to essentials, and adhere to Albert Einstein's dictum that "Everything should be as simple as possible, but not simpler."

Systematically, the chapters progress from biometry with focus on axial length, corneal power, and intraocular lens position; to formulas for power calculation including formula selection and personalization; and on to clinical factors such as patient preference, concurrent ocular conditions, and "intraocular lens surprises." Each segment is presented authoritatively with the strength of extensive experience.

The subject of preoperative intraocular lens power calculation is more important than ever before, because of the requirement for accuracy related to multifocal, accommodative, and toric intraocular lenses. To this must be added the complexities of intraocular lens power calculation after corneal refractive surgery and the increased expectations of patients.

As a basic text and a reference for use in special circumstances, IOL Power is of distinct value to ophthalmologists, para-ophthalmic personnel, and manufacturers of intraocular lens devices. Admirably, Dr. Hoffer and co-authors succeed in the stated objective of providing information to improve the accuracy of intraocular lens power calculations in ophthalmic surgery.

Bradley R. Straatsma, MD, JD
Chairman Emeritus
Jules Stein Eye Institute
University of California, Los Angeles

Basics and Accurate Biometry

Section I: Introduction

Kenneth J. Hoffer, MD (Fig. 1-1)

Figure 1-1. Kenneth J. Hoffer, MD

This book has been written to provide all the latest information available regarding the calculation of intraocular lens (IOL) power. It is formatted along the lines of the IOL power courses I have taught at the annual meetings of the American Academy of Ophthalmology (AAO) and the American Society of Cataract & Refractive Surgery (ASCRS) over the past 36 years (Fig. 1-2). After years of presenting these courses, it seemed this was a good time to put this material into a textbook to increase the availability of the information to everyone.

When I performed the first ultrasound A-scan for IOL power in the United States in 1974, I also wrote the original Hoffer formula. I was unable to get it published in American journals because it was concerning IOLs and I was just two years out of my residency and completely unknown. The first time I was able to publish anything was in 1975[1] (Fig. 1-3), after I started the American Intra-Ocular Implant Society (AIOISJ), which is now the *Journal of Cataract & Refractive Surgery* (*JCRS*). It took me 7 years to finally get the Hoffer formula[2] published in 1981. A similar thing happened when I performed a study that showed that the ultimate position of the IOL was directly proportional to the axial length (AL) of the eye[3] (Fig. 1-4). This led me to propose a

4   Chapter 1

Figure 1-2. Title slide for IOL Power courses given at AAO and ASCRS since 1975.

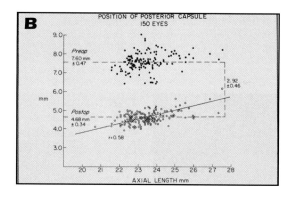

Figure 1-3. First published paper on IOL power calculation in the US in AIOISJ, 1975.

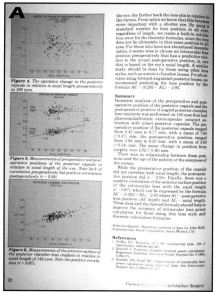

Figure 1-4. First published paper establishing direct relationship between IOL position and the AL of the eye, 1975.

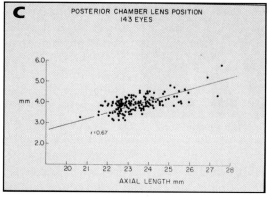

Section I: Introduction 5

Figure 1-5. Collage of textbooks containing chapters on IOL power by the author.

Figure 1-6. Kwitko-Kelman book containing chapter on the history of IOL power calculation in North America.

Figure 1-7. IOL Power Club formed in San Sebastian, Spain on September 5, 2005.

formula to better predict the IOL position using AL. Just as I was preparing to submit it for publication, the Chairman of my residency department at Kresge Eye Institute strongly requested me to submit something to this new, short-lived journal he had been asked to edit. I thus submitted it to him and it was published in a short-lived journal that is difficult to reference.

Over the past 35 years, I have been asked to write a textbook but have never been able to get to it. On the other hand, I have written chapters on this subject in more than 50 textbooks of others (Fig. 1-5). I was asked to write the history of IOL power calculation in North America for the book on the history of modern cataract surgery by Kwitko and Kelman (Fig. 1-6).

I wish to thank my colleagues, the fellow members in the IOL Power Club (IPC) (Fig. 1-7), and others who have graciously taken their time to make contributions based on their

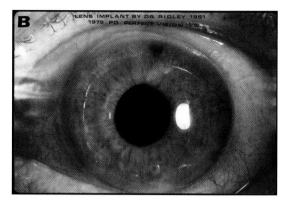

Figure 1-8. (A) Sir Harold Ridley, MD, London, UK. (B) Eye with a Ridley implant with UCVA of 20/20, implanted by Ridley in 1952, photographed in 1979 (photo by Hoffer). (C) The author (center) with Lady Elizabeth and Sir Harold Ridley (right) in 1999. (Photo by Marcia Hoffer.)

expertise. The IPC was formed in 2005 in San Sebastian, Spain to foster camaraderie, communication, and scientific interaction in this subject by those who have labored in this field for many years. The IPC was founded by Jaime Aramberri of San Sebastian, Spain; Wolfgang Haigis, PhD, of Würzburg, Germany; Sverker Norrby, PhD, of Groningen, Holland; Thomas Olsen, MD, of Århus, Denmark; H. John Shammas, MD of Lynwood, CA, USA; and myself. The joint efforts of these pillars of this field have led to new communication and many new ideas.

Since Sir Harold Ridley (Fig. 1-8A and C) experienced a 21 diopter (D) "surprise" in lens power calculation on his first two cases in 1949-1950, we have been seeking ways to calculate intraocular lens (IOL) power with greater accuracy. Fig. 1-8B (photo by author in 1979) proves he ultimately worked this out, since this is an eye with a Ridley posterior chamber (PC) implant with uncorrected visual acuity (UCVA) of 20/20, implanted by Ridley in 1952.

This science is rather dry and does not stimulate great interest on the part of the majority of cataract surgeons. To make the subject more interesting it might be advantageous to break it down into its component parts. The three major components of IOL power calculation are Biometry, Formulas, and *Clinical Variables*. Biometry can be divided into its components needed to calculate IOL power: the *AL*, the *Corneal Power (K)*, and the *IOL Position (ELP)*. The subject of Formulas involves their *Generations*, their *Usage*, and their *Personalization*. Clinical Variables deals with the subjects of *Patient Needs and Desires, Special Circumstances,* and *Problems and Errors*. This book will present all the information necessary to fully understand all the aspects of calculating IOL powers.

- Biometry
  - Axial length (AL)
  - Corneal power (K)
  - IOL position (ELP)
- Formulas
  - Generations
  - Usage
  - Personalization
- Clinical Variables
  - Patient Needs and Desires
  - Special Circumstances
  - Problems and Errors

We recommend for references in this subject the textbooks by H. John Shammas[4] and Sandra Frazier-Byrne.[5]

References

1. Hoffer KJ. Mathematics and computers in intraocular lens calculation. *Am Intra-Ocular Implant Soc J.* 1975;1(1):3.
2. Hoffer KJ. Intraocular lens calculation: The problem of the short eye {Hoffer formula}. *Ophthalmic Surgery.* 1981;12:269-272.
3. Hoffer KJ. The effect of axial length on posterior chamber lenses and posterior capsule position. *Current Concepts in Ophthalmic Surgery.* 1984;1:20-22.
4. Shammas HJ, ed. *Intraocular Lens Power Calculations.* Thorofare, NJ: SLACK Incorporated; 2003.
5. Byrne SF. *A-scan Axial Eye Length Measurements.* Mars Hill, NC: Grove Park Publishers; 1995.

Axial Length: Ultrasound
Background

Kenneth J. Hoffer, MD

When the human lens is replaced with an IOL (Fig. 2-1), the optical situation becomes a two-lens system (cornea and IOL) projecting an image onto the retina (macula). The distance between the two lenses (Estimated Lens Position, or ELP) affects the refraction, as does the distance between the two-lens system and the macula (Y). ELP is defined as the distance from the anterior surface (vertex) of the cornea to the effective principle plane of the IOL in the visual axis. Y is defined as the distance from that principle plane of the IOL to the photoreceptors of the macula in the visual axis. It is easy to see that ELP + Y is equal to the visual axis AL of the eye. Therefore, knowing the ELP and the AL will allow the calculation of Y (Y = AL − ELP).

Also to calculate the IOL power (P), we must know the optical effective power of the cornea (K) as well as the vergence of the light rays entering the cornea (refractive error, or R). For emmetropia, R is zero. The relationship of the factors P, AL, K, ELP, and R are such that a formula can be written to describe it. The first such formula for IOL power was published in Russia by Fyodorov and Kolonko[2] in 1967. Knowing the values of any four of these values will allow for the calculation of the 5th.

Originally it was thought that it was good enough to get the patient back to the refractive error they had prior to the cataract. It became standard to use the same power for all eyes to yield this result. As lens implantation became more popular in the US, a standard 18.0 D Binkhorst prepupillary lens was used. Some thought that they could adjust this power up or down depending upon the previous refractive error, and charts were developed to aim closer to emmetropia. The first IOL formulas became available in the late 1960s and early 1970s but they required the need to measure the AL of the eye and the power of the cornea (K).

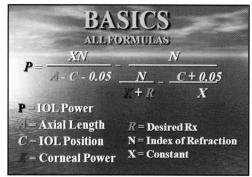

Figure 2-1. Basics of all formula's calculation of IOL power.

Figure 2-2. Jan Worst MD, Holland.

Figure 2-3. Karl Ossoinig, MD, Iowa, the father of accurate immersion A-scan ultrasound for axial length measurement.

Early attempts were made to measure the AL using ultrasound in Russia and Holland. In the early 1970s, Jan Worst of Groningen, Holland (Fig. 2-2) used a small A-scan ultrasound with the Colenbrander[1] formula that was also used by several of the early Dutch IOL implanters. I became aware of this and, having knowledge of the work done by Ossoinig[2] (Fig. 2-3) to develop the Kretz 7200MA A-scan ultrasound in Vienna, Austria, decided to use this instrument to perform the first such AL measurements in the US in April 1974.[3] Ossoinig had improved the accuracy of AL measurement with a precise immersion cup technique and a specific calibration of the instrument.

Because the technique required photographing the screen and making measurements of the Polaroid photographs with precise calipers, the process was quite tedious. I trained the first American A-scan technician (Donald Allen) and then performed a study that showed that technician accuracy was equal to that of a physician.

I then set out to persuade Sonometrics, Inc (Boston, MA) to produce an A-scan that was specific for IOL power and produce an automatic readout of the AL. In 1975, they produced the Sonometrics DBR-100 (Fig. 2-4), which used an applanation cone applied directly to the cornea and a fixation light as I had directed. The instrument read out the AL digitally without need of photos and calipers. Unfortunately, though it became popu-

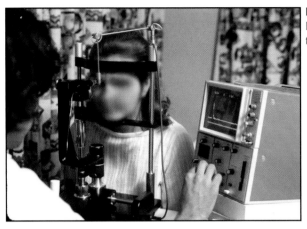

Figure 2-4. Original 1975 Sonometrics DBR-100 A-scan ultrasound using the contact applanation method.

lar throughout the world, it used an applanation contact method, later proven by us to be inferior to immersion.

I compared the results of this instrument with the standard immersion technique and discovered that the applanation technique led to an unpredictable shortening of the reading from 0.25 to 0.33 mm, which could not be offset by a correction factor. After informing others of this fact,[4] Shammas[5] and others[6] repeated my study and validated this finding with similar results. In the meantime, many adopted the instrument and applanation technique and it became the standard in the US, as it was considered much easier than the Kretz unit and the Ossoinig immersion technique.

Fyodorov[7] published the first IOL power formula in 1967. I[8] developed a formula, based upon the Colenbrander formula, in 1974 and used it with the immersion technique. The accuracy of the results led to a standard of aiming for emmetropia in all cataract patients. Courses and lectures over the ensuing years gradually developed a standard of not using just one IOL power for all eyes, but rather individually calculating an IOL for each patient. Ultrasound instruments were continuously developed over the next 20 years. Everything changed in 1999 when Zeiss (Jena, Germany) introduced the first instrument that uses a laser to measure the AL. The IOLMaster was specifically set so that it matched the results with the immersion ultrasound technique and not applanation. This instrument has now been shown to be easier to use, more accurate, and more reproducible than any other instrument or technique. In 2009, Haag-Streit (Koeniz, Switzerland) produced a similar unit they call the LENSTAR LS900.

In the 1980s, because of inaccuracies due to the applanation method, many were looking for ways to improve IOL calculation. This led to the development of regression formulas, which soon coalesced into the SRK formula.[9] Because it was so easy to use, it became rapidly popular worldwide. I first showed that it was the leading cause of IOL removals because of IOL power error.[7] Holladay's introduction of his formula[10] that improved the prediction of the ELP in 1990 helped lead many back to the use of theoretic formulas. This led to the development of the SRK/T[11] (the "T" stands for Theoretic) formula by Retzlaff, the Hoffer Q[12] formula in the early 1990s, and the Haigis[13] formula in 2000.

Modern IOL power calculation has been the product of many small steps and a few steps backwards (regression formulas), but today we can assure patients that their chance of getting an improper IOL power is quite small.

References

1. Colenbrander MC. Calculation of the power of an iris clip lens for distance vision. *Br J Ophthalmol.* 1973;57(10):735–740.
2. Ossoinig KC. Standardized echography: Basic principles, clinical applications, and results. *Int Ophthalmol Clin.* 1979;19(4):127–210. Review. No abstract available.
3. Hoffer KJ. The history of IOL power calculation in North America. In: Kwitko ML, Kelman CD, eds. *The History of Modern Cataract Surgery.* The Hague, Netherlands: Kugler Publications; 1998;193–208.
4. Comparison of Kretz 7200 MA, Sonometrics DBR, and Storz Oculometer for Lens Calculation, First U.S. Intraocular Lens Symposium. American Intra-Ocular Implant Society (AIOIS), Los Angeles, CA, March 1978.
5. Shammas HJ. A comparison of immersion and contact techniques for axial length measurements. *J Am Intraocul Implant Soc.* 1984;10(4):444–447.
6. Schelenz J, Kammann J. Comparison of contact and immersion techniques for axial measurement and implant power calculation. *J Cataract Refract Surg.* 1989;15(4):425–428.
7. Fyodorov SN, Kolonko AI. Estimation of optical power of the intraocular lens. Vestnik Oftalmologic (Moscow). 1967;4:27.
8. Hoffer KJ. Intraocular lens calculation: The problem of the short eye. *Ophthalmic Surg.* 1981;12(4):269–272.
9. Sanders D, Retzlaff J, Kraff M, et al. Comparison of the accuracy of the Binkhorst, Colenbrander, and SRK implant power prediction formulas. *J Am Intraocul Implant Soc.* 1981;7(4):337–340.
10. Holladay JT, Prager TC, Chandler TY, et al. A three-part system for refining intraocular lens power calculations. *J Cataract Refract Surg.* 1988;14(1):17–24.
11. Retzlaff J, Sanders DR, Kraff MC. Development of the SRK/T intraocular lens implant power calculation formula. *J Cataract Refract Surg.* 1990;16(3):333–340. Erratum in: 1990;16(4):528.
12. Hoffer KJ. The Hoffer Q formula: A comparison of theoretic and regression formulas. *J Cataract Refract Surg.* 1993;19(6):700–712. Erratum in: 1994;20(6):677 and 2007;33(12):2–3.
13. Haigis W. The Haigis Formula. In: Shammas, HJ, ed. *Intraocular Lens Power Calculations.* Thorofare, NJ: SLACK Incorporated; 2003:41–57.

3

Immersion Versus Applanation Ultrasound

Kenneth J. Hoffer, MD

The immersion technique of Ossoinig[1] has been shown to be more accurate than the standard applanation technique in several studies[2-6] over the past 3 decades. They report a mean average shortening of the AL of 0.21 mm (range from 0.11 to 0.36 mm) using applanation compared to immersion, and the longer the eye the greater the shortening[4] (Table 3-1).

Theoretically, if this iatrogenic error was consistent it could be compensated for by the addition of a simple constant or by formula personalization, but this is not possible since the error varies so much from eye to eye.

Arguments against immersion are that it is expensive, time-consuming, messy, and requires the patient to be supine. On the contrary, the examination can be performed very easily in a standard ophthalmic examination chair (Fig. 3-1A) reclined back at a 45° angle with the headrest set back so that the patient's AL is perpendicular to the floor (Fig. 3-1B).

To maintain a non-leaking fluid bath in the Ossoinig scleral shell (Fig. 3-2A) (Hansen Ophthalmic Development Labs, Coralville, IA, www.HansenLab.com), we use a 50/50 dilution of 2.5% hydroxypropyl methycellulose (Goniosol, CIBA Vision Ophthalmics, Atlanta, GA) in Dacriose solution (Fig. 3-2B through D). Once the eye is anesthetized topically, the scleral shell is gently placed between the lids and filled three-quarters full with the solution. Any air bubbles should be vacuumed with a short silicone tube attached to a syringe. The latter can also be used to remove the solution at the completion of the procedure. The ultrasound probe is placed into the solution and positioned parallel to the

Table 3-1.

Applanation vs Immersion Studies in the Literature Since 1981

Report by	# Eyes	Applanation US	Immersion US	Shortening
Hoffer 1981	28	22.77	23.10	0.33
Shammas 1984	180	23.28	23.52	0.24
Artaria 1986	131	23.13	23.44	0.31
Schelenz and	46 (<23.3)	22.39	22.59	0.20
Kammann 1989	54 (>23.3)	24.06	24.38	0.32
Olsen 1989	60	23.35	23.49	0.14
Watson and Armstrong 1999	225	23.24	23.55	0.11
TOTAL	724			0.21

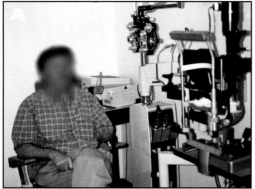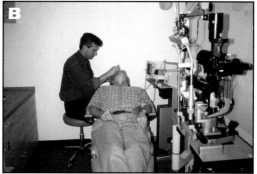

Figure 3-1. Patient sitting (A) and then reclined (B) in standard ophthalmic exam chair for immersion technique.

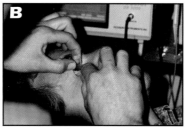

Figure 3-2. Immersion setup. (A) Gonisol cut 50% with Dacriose, (B) Ossoinig cup placed between lids, (C) solution fills cup, (D) ultrasound probe placed in solution and parallel to the eye's axis.

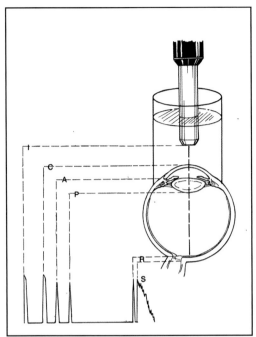

Figure 3-3. Axiality is indicated by spike patterns on the oscilloscope screen as the probe position is adjusted.

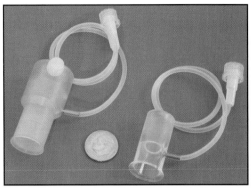

Figure 3-4. Prager (left) and Kohn shells are an alternative immersion technique.

axis of the eye. Axiality is judged by watching for the correct spike patterns on the oscilloscope screen as the probe position is adjusted (Fig. 3-3).

The first spike is the echo emission from the end of the probe, the second is the double-spike from the cornea, the third is the anterior of the lens, the fourth is the posterior of the lens, the fifth is the retinal spike, and the remaining are the spikes from the sclera and orbital tissues. It is important to adjust the sound intensity so that the spikes can be seen to change their height. Then the probe is adjusted so that the corneal and retinal spikes are about equally high and straight. Then the reading should be taken.

Many find the Prager or Kohn shells (ESI, Inc, Plymouth, MN) (Fig. 3-4) easier to use for immersion. I have no experience using it; therefore see Chapter 7 on its use.

NOTE: If the AL is very difficult to obtain and the eye appears to have a length greater than 25 mm, suspect a STAPHYLOMA. Shammas recommends direct ophthalmoscopy (with patient fixating on cross-hair target); measure the distance from the target (macula) to the edge of the optic nerve (in disc diameters). B-scan exam is then performed to measure the AL at that distance from the edge of the optic nerve shadow (personal communication) (see Chapter 27).

NOTE: When measuring an eye containing an IOL, ignore multiple reduplication echoes noted in the vitreous space that are caused by the IOL.

NOTE: If planning silicone oil injection into the vitreous space, perform an accurate AL measurement before doing so, and make this information available to the patient. It is practically impossible to measure a silicone oil eye. If your US instrument allows you to do so, set the vitreous gate to the corresponding sound velocity of the silicone oil (either 1000 cSt or 5000 cSt. Otherwise, using a velocity of 1000 m/s may not achieve the desired accuracy. The Zeiss IOLMaster is the best way to get a measurement in silicone oil-filled eyes (see Chapter 11).

NOTE: *Measuring the AL of BOTH eyes is prudent and customary.*

Always measure AL to the nearest hundredth of a millimeter and record it carefully. Errors in AL are the most significant and amount to ~2.5 D/mm in IOL power. But, it is important to be aware that this error drops to ~1.75 D/mm in very long eyes (30 mm) and jumps to ~3.75 D/mm in very short eyes (20 mm). Greater care must be taken in measuring short eyes.

References

1. Ossoinig KC. Standardized echography: Basic principles, clinical applications, and results. *Int Ophthalmol Clin.* 1979;19(4):127–210.
2. Shammas HJF. A comparison of immersion and contact techniques for axial length measurements. *J Am Intraocul Implant Soc.* 1984;10(4):444–447.
3. Artaria LG. Axial length measurements with different ultrasound devices. *Klin Monast Augenheilkd.* 1986;188:492-494.
4. Schelenz J, Kammann J. Comparison of contact and immersion techniques for axial measurement and implant power calculation. *J Cataract Refract Surg.* 1989;15(4):425–428.
5. Olsen T, Nielsen PJ. Immersion versus contact technique in the measurement of axial length by ultrasound. *Acta Ophthalmol (Copenh).* 1989;67(1):101–102.
6. Watson A, Armstrong R. Contact or immersion technique for axial length measurement? *Aust N Z J Ophthalmol.* 1999;27(1):49–51.

4

Ultrasound Velocities for Axial Length Measurement

Kenneth J. Hoffer, MD

Measuring the AL of the eye using an A-scan is dependent upon the sound velocity the instrument is set at for the measurement. Some instruments use an average velocity for the entire eye while others use individual velocities for each part of the eye. It would be valuable for you to find out what method your instrument uses.

The ultrasound velocities for the various parts of the eye and intraocular lens materials are shown in Table 4-1 and the average pseudophakic velocities that the author[1] calculated in 1974 are shown in Table 4-2. The effect of AL error is 2.35 D for every mm in a 23.5 mm eye, but drops to 1.75 D/mm in a 30 mm eye. It rises dramatically to 3.75 D/mm in a 20 mm eye, which is the reason short eyes are the problem eyes in IOL power (Fig. 4-1).

To calculate the average sound velocity in a normal PHAKIC eye of 23.5 mm, the average thickness of the cornea (0.55) and lens (4.63) must be used to individually calculate the time it takes the sound wave to travel through each of them. This is done by dividing them individually by their respective velocity (1641 m/sec) (Fig. 4-2). The remaining fluid parts of the eye (23.5 - 0.55 - 4.63) must then be divided by the fluid velocity (1532 m/sec). The total time it takes for the sound to traverse the entire eye is the sum of the "solid" time and the "liquid" time. We can use the formula Velocity = Distance/Time (V = d/t). Since the eye has a given AL of 23.5 mm, that value is divided by the total time (15.1148 m/sec) yielding an average velocity of 1555 m/sec. If this same exercise is performed on a sample long eye and short eye (Fig. 4-3) the calculations reveal that the average velocity is not the same as for a normal length eye. A 30 mm eye thus has a velocity of 1549 m/sec and a 20 mm eye has a velocity of 1561 m/sec.

Table 4-1.

ULTRASOUND VELOCITIES (METERS/SECOND) AT BODY TEMPERATURE[2]

Substance	Velocity (m/sec)
Cornea and lens	1641
Aqueous and vitreous	1532
PMMA IOL	2660
Silicone IOL	980
Acrylic IOL	2026
Glass IOL	6040
Silicone oil	987

Table 4-2.

HOFFER CALCULATED AVERAGE SOUND SPEEDS (METERS/SECOND) FOR VARIOUS CONDITIONS OF A 23.5-MM EYE[1]

Eye Status	Velocity (m/sec)
Phakic eye	1555
Aphakic eye	1534
PMMA pseudophakic	1556
Silicone pseudophakic	1476
Acrylic pseudophakic	1549
Glass pseudophakic	1549
Phakic silicone oil	1139
Aphakic silicone oil	1052

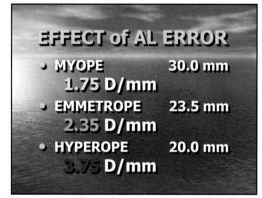

Figure 4-1. Effect of an error in axial length on resultant refractive error.

Figure 4-2. Calculations to determine the average sound velocity of a normal size phakic eye.

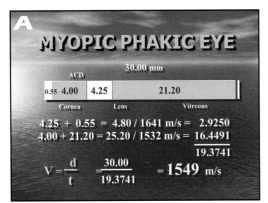 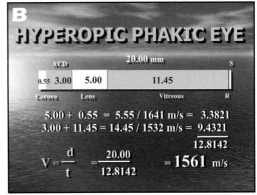

Figure 4-3. Average velocity for long (A) and short (B) eyes.

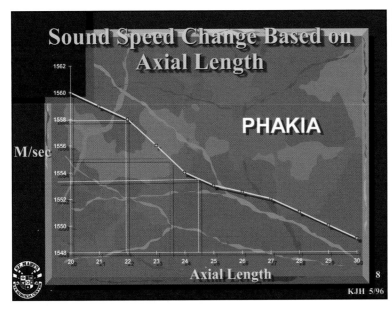

Figure 4-4. Graph of the average velocity (Y-axis) based on the AL (X-axis).

Graphing the average velocity against AL reveals an inverse relationship to the AL (Fig. 4-4). As can be seen, the slope is steeper in short eyes than it is in long eyes.

Now if we do the same exercise on an example normal length **aphakic** eye (Fig. 4-5A,) we obtain a drop in average velocity from 1555 to 1534 m/sec, which is totally due to the loss of the crystalline lens. Calculating the same for long and short aphakic eyes (Fig. 4-5B) reveals that AL does not seem to affect the average velocity. This tells us that it is the lens in the eye that causes this average velocity variation.

It is important to perform this calculation on **pseudophakic** eyes with IOLs of different materials. We can perform this calculation on eyes of normal AL and thus for a PMMA lens (Fig. 4-6A) of average thickness of 0.74 mm we obtain an average velocity of 1555 m/sec, which is the same as a phakic eye. If we replace the lens with one of Collamer (Staar Surgical Company, Monrovia, CA) (Fig. 4-6B) of average thickness of 0.74 mm, the average velocity drops to 1540 m/sec.

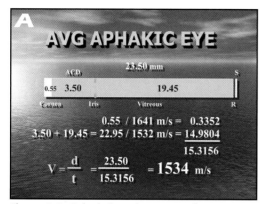
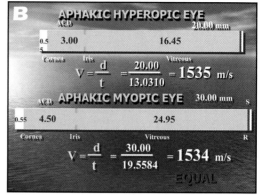

Figure 4-5. Calculation of average velocities in a normal size aphakic eye (A) and long and short aphakic eyes (B).

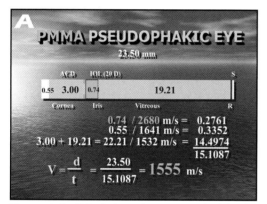
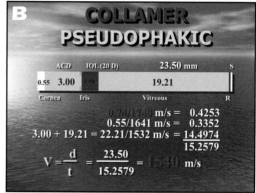

Figure 4-6. Calculation of average velocities in a normal size PMMA pseudophakic eye (A) and a Collamer pseudophakic eye (B).

Figure 4-7. Calculation of average velocity in a normal size silicone pseudophakic eye.

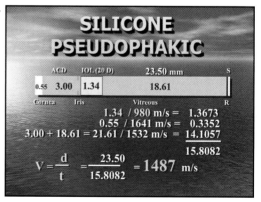

Repeating this for a normal eye with a silicone lens (Fig. 4-7) of average thickness 1.34 mm which has a sound velocity of only 980 m/sec, we see a dramatic drop of the average eye velocity to 1487 m/sec. This will have a substantive effect on the IOL power calculation leading to clinically significant errors if 1555 m/sec is used to measure the eye.

NOTE: *Measuring an eye containing a silicone IOL with a standard phakic velocity (1555 m/sec) can amount to an error of 3 to 4 D.*

CORRECTING AN ERROR IN VELOCITY

If an eye has been measured using the wrong velocity, it can be easily corrected without remeasuring the eye by using the following formula:

$$AL_C = (AL_M) \times (V_C) \div V_M$$

where AL = axial length, V = ultrasound velocity, C = correct and M = measured.

This is because the ultrasound instrument does not measure length or distance (d) directly. Instead it measures the time (t) it takes the sound to traverse the eye and converts it to a linear value using the velocity (V) formula where d = V x t.

OPTIONAL CALF METHOD

Holladay[3] published an optional method to measure the AL that attempts to decrease the error inherent in changes in average velocity due to the length of the eye. All eyes, regardless of status, are measured at a velocity of 1532 m/sec (as if the eye was a bag of water) and to this value is added the Corrected AL Factor (CALF). The CALF value represents the thickness of a lens in the eye whether it is the crystalline lens or an IOL(s). The formula for the CALF of any lens (including the cornea) is:

$$CALF = T_L \times (1 - 1532/V_L)$$

where T_L = the axial thickness of the lens and V_L = the sound velocity through that lens.

Holladay computes the thickness of the human cataractous lens using:

$$T_L = 4 + Age/100$$

and the sound velocity through the cataract using:

$$V_L = 1659 - [(Age - 10)/2]$$

Substituting these two formulas into the CALF formula above, the CALF formula for the crystalline lens yields:

$$CALF = \left[4 + \frac{Age}{100}\right] \times \left[1 - \frac{1532}{\left(1659 - \left(\frac{Age - 10}{2}\right)\right)}\right]$$

CALF formula is dependent only upon the age of the patient.

As can be seen, the CALF for the cataractous lens is calculated using only the age of the patient. Holladay recommends using a CALF value of 0.28 (value for 70 year-old) for all ages because the value for a 1-year-old is 0.306 and that for a 100-year old is 0.224. The maximum error in CALF for those younger than 70 is 0.026 (~0.07 D) and for those older than 70 it is 0.056 (~0.14 D). The reasoning behind this method is that, if an "average" eye velocity is incorrect, it affects the entire AL measurement. However, if the estimate of the CALF value is wrong, it only affects a small percentage of the overall AL, ie, only the lens portion.

> **Table 4-3.**
>
> ### FORMULAS FOR CALCULATING BIOMETRIC PARAMETERS
>
> A. CALF factors for pseudophakic eyes (using CALF = $T_L \times (1 - 1532/V_L)$) where V_L = the sound velocity for the IOL material in the eye:
>   - $CALF_{PMMA} = T_L \times (1 - 1532/2660) = +0.424 \times T_L$
>   - $CALF_{SILICONE} = T_L \times (1 - 1532/980) = -0.563 \times T_L$
>   - $CALF_{ACRYLIC} = T_L \times (1 - 1532/2026) = +0.243 \times T_L$
>
> B. The correction for the cornea:
>   - $CALF_{CORNEA} = T_C \times (1 - 1532/1641) = 0.55 \times (0.066423) = 0.037$
>
> C. Knowing the thickness of the implanted IOL (which can be obtained from the manufacturer) the following formulas can be used:
>   - PMMA Eye $AL = AL_{1532} + 0.424 \times TL + 0.037$
>   - Silicone Eye $AL = AL_{1532} - 0.563 \times TL + 0.037$
>   - Acrylic Eye $AL = AL_{1532} + 0.243 \times TL + 0.037$
>   - Piggyback IOLs $AL = AL_{1532} + T1 \times (1 - 1532/V1) + T2 \times (1 - 1532/V2) + 0.037$ where T1 and T2 are the thickness and V1 and V2 are the velocity of each IOL.

CORNEA MUST BE CONSIDERED

His formulation, however, ignores the factor of the corneal thickness (0.55 mm). To correct this, I recommend using a CALF of 0.32 (0.28 + 0.037). The 0.37 correction for the cornea is calculated in Table 4-3B.

A similar method can be used for pseudophakic eyes using CALF = $T_L \times (1-1532/V_L)$ and the known V_L for each IOL material (Table 4-3A). Knowing the thickness of the implanted IOL, the formulas in Table 4-3C can be used. If the IOL thickness cannot be obtained, Holladay[3] published a table to use based on the power of the IOL. The resulting formulas based on IOL material are shown in Fig. 4-8. The AL of an eye containing two IOLs of different materials can be obtained using the formula in Table 4-3C.

BIPHAKIC EYES (PHAKIC EYES WITH A PHAKIC IOL)

A phakic eye with a phakic IOL implanted (biphakic) may develop a cataract. The problem here is eliminating the effect of the sound velocity through the phakic lens when measuring the AL using ultrasound. The A-scan assumed that sound travels through the phakic refractive lens (PRL), for example, at 1555 m/sec; however, sound travels through the PRL at 980 m/sec. To calculate the true AL of this eye, we must calculate and subtract the erroneous (E) distance of the PRL and add back the true (T) distance (Fig. 4-9):

$$AL + T - E$$

where $E = T \times 1555/980$.

The phakic IOL thickness can be obtained from the publications[4,5] which have tables showing the phakic IOL central thickness based on their dioptric power for each phakic IOL on the market today.

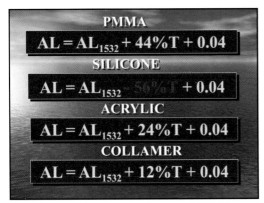

Figure 4-8. CALF formulas for pseudophakic eyes containing different materials.

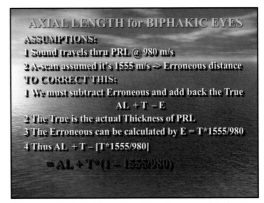

Figure 4-9. Axial length with ultrasound in biphakic eyes.

Then the
$$AL + T - E = AL + T - (T \times 1555/980) = AL + T \times (1 - 1555/980).$$

The author proposed a method[4] to correct for this potential error by using the following resultant formula:
$$AL_{CORRECTED} = AL_{1555} + C * T$$

where AL_{1555} = the measured AL of the eye at sound velocity of 1555 m/sec, T = the central thickness of the phakic IOL and C = the material specific correction factor of +0.42 for PMMA, -0.59 for silicone, +0.11 for collamer, and +0.23 for acrylic.

References

1. Hoffer KJ. Ultrasound velocities for axial length measurement. *J Cataract Refract Surg.* 1994;20(5):554–562.
2. Mark HF, Bikales N, Overberg CG, Menges, Kroschwitz G JI. *Encyclopedia of Polymer Science and Engineering,* Volume 1, 2nd ed. New York: Wiley and Sons. 1985;147–149.
3. Holladay JT. Standardizing constants for ultrasonic biometry, keratometry, and intraocular lens power calculations. *J Cataract Refract Surg.* 1997;23(9):1356–1370.
4. Hoffer KJ. Ultrasound axial length measurement in biphakic eyes. *J Cataract Refract Surg.* 2003;29(4):961–965.
5. Hoffer KJ. Addendum to ultrasound axial length measurement in biphakic eyes: Factors for Alcon L12500–L14000 anterior chamber phakic IOLs. *J Cataract Refract Surg.* 2007;33(5):751–752.

Suggested Reading

Holladay JT, Prager TC. Accurate ultrasonic biometry in pseudophakia. *Amer J Ophthalmol.* 1989;107(2):189–190.

A-Scan Biometry: Immersion Technique

H. John Shammas, MD

The immersion technique described herein can be used with any ultrasound unit equipped with a solid A-scan probe and mobile electronic gates.[1-5]

- The patient is placed in a supine position on a flat examination table or in a reclining examination chair and a drop of local anesthetic is instilled in both eyes.
- An Ossoinig scleral shell is applied between the lids. The most commonly used scleral shells are the Hansen shells, the Prager shell, and the Kohn shell (Fig. 5-1).
- The Ossoinig shell is filled with gonioscopic solution. Methylcellulose 1% is preferred over the 2.5% concentration (which is too thick) and over saline solutions (too liquid). The solution should be free of air bubbles; the presence of bubbles causes variations in the speed of sound and is responsible for noise formation within the ultrasound pattern. The easiest way to avoid bubbles is to remove the bottle's nipple and to pour the solution in the cup. If bubbles do form within the solution, they are removed with a syringe, and, if unsuccessful, the cup has to be emptied, cleaned, repositioned and refilled with gonioscopic solution.
- The Prager and the Kohn shells are designed to hold the probe tightly and allow a better fit on the eye. Because of this tight fit, the coupling fluid used in these shells does not have to be methylcellulose; instead, balanced salt solution or artificial tears could be used.

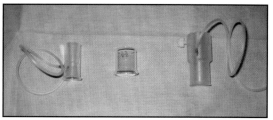

Figure 5-1. The Kohn, Ossoinig (Hansen), and Prager scleral shells are displayed from left to right.

Figure 5-2. The ultrasound probe is immersed in the solution keeping it 5 to 10 mm away from the cornea.

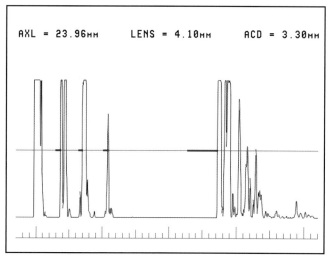

Figure 5-3. Ultrasound display of the different echospikes during immersion A-scan biometry, identifying from left to right: the initial spike, the anterior and posterior corneal surfaces, the anterior and posterior lens surfaces, the retina, sclera, and orbital tissues.

- The ultrasound probe is inserted in the Kohn shell that keeps the tip 5 to 8 mm away from the cornea (Fig. 5-2). The patient is asked to look, with the fellow eye, at a fixation point placed at the ceiling. Gently move the probe until it is properly aligned with the optical axis of the eye and an acceptable A-scan echogram is displayed on the screen.

The A-scan pattern of a normal phakic eye examined with an immersion technique displays the following echo spikes from left to right (Fig. 5-3):
- The initial spike is produced at the tip of the probe. It has no clinical significance.
- The corneal spike is double-peaked, representing the anterior and posterior surfaces of the cornea.
- The anterior lens spike is generated from the anterior surface of the lens.
- The posterior lens spike is generated from the posterior surface of the lens.
- The retinal spike is generated from the anterior surface of the retina. It is straight, highly-reflective, and tall whenever the ultrasound beam is perpendicular to the retina (as it should be during AL measurement).

- The scleral spike is another highly-reflective spike generated from the scleral surface right behind the retinal spike, and should not be confused with it.
- The orbital tissues create low reflective spikes behind the scleral spike.

Many older biometers give the readings directly in millimeters using an average sound velocity of 1550 to 1555 m/s.[6,7] The most accurate velocity to use is noted in Table 4-2 of the Ultrasound Velocities chapter. Most modern biometers use individual velocities for each of the eye's components (cornea, aqueous, lens, and vitreous), but it is important to make sure the values are set to the appropriate ones shown in Table 4-1 of the Ultrasound Velocities chapter. It helps to replace older US units with more modern ones or an optical measuring device such as the IOLMaster or LENSTAR LS900.

- The thickness of the cornea is measured using a velocity of 1641 m/sec.
- The anterior aqueous depth is measured between the posterior corneal surface and the anterior lens surface using a velocity of 1532 m/sec.
- The lens thickness is measured between the anterior lens surface and the posterior lens surface using a velocity of 1641 m/sec. Actually 1640.5 m/sec is the calculated sound velocity in the normal crystalline lens. The sound velocity varies in cataractous eyes with a slower velocity (average 1590 m/sec) in the intumescent cataracts due to their high water content, and a higher velocity in the posterior capsular cataracts. In most cases of nuclear sclerosis with or without subcapsular changes, the sound velocity averages 1641 m/sec.
- The vitreous cavity's depth is measured between the posterior lens surface and the anterior surface of the retina using a velocity of 1532 m/sec.

References

1. Byrne SF. Standardized echography, Part I: A-scan examination procedures. *Int Ophthalmol Clin*. 1979;19(4):267–281. No abstract available.
2. Ossoinig KC. Standardized echography: Basic principles, clinical applications and results. *Int Ophthal Clin*. 1979;19(4):127–210.
3. Shammas HJ. Axial length measurement and its relation to intraocular lens power calculations. *J Am Intraocul Implant Soc*. 1982;8(4):346–349.
4. Shammas HJ. Manual versus electronic measurement of the axial length. In: Hillman JS, LeMay MM, eds. *Ultrasonography in Ophthalmology. Proceedings of the 1982 Ninth SIDUO Congress*. The Hague: Dr. W. Junk Publishers; 1983:225–229.
5. Shammas HJ. A-scan biometry of 1000 cataractous eyes. In: Ossoining KC, ed. *Ophthalmic Echography. Proceedings of the 10th SIDUO Congress*. The Hague: Dr. W. Junk Publishers; 1987:48,57–63.
6. Oksala A, Lehtinen A. Measurement of the velocity of sound in some parts of the eye. *Acta Ophthalmol*. 1958;36(4):633–639.
7. Hoffer KJ. Ultrasound velocities for axial length measurement. *J Cataract Refract Surg*. 1994;20(5):554–562.
8. Coleman DJ, Lizzi FL, Franzen LA, Abramson DH. A determination of the velocity of ultrasound in cataractous lenses. *Bibl Ophthalmol*. 1975;83:246–251.

A-Scan Biometry: Avoiding Pitfalls With Immersion Technique

Karl Ossoinig, MD

Historical Background

Early in the 1960s, the Swedish ophthalmologist F. Janson,[1,2] through his extensive experimental and clinical work, introduced precise and accurate immersion AL measurements. He also gave us the exact sound velocities within the clear lens and the vitreous body as well as the anterior chamber fluid. Hoffer introduced immersion A-scan for IOL power calculation in the USA in 1974 and stimulated Sonometrics to produce a stand-alone unit that introduced applanation contact A-scan. He later proved (see Chapter 1) that applanation shortened the eye compared to immersion and abandoned its use. This method however gained immense popularity because it was considered an easier and faster method. It spread rapidly and widely for mainly two reasons:
- It required no additional space in ophthalmic offices to recline a patient as needed for the immersion method.
- It was easier to learn and quicker to perform.

Why Immersion Method

Over the past decade the increasing interest in the immersion method is due to the following reasons:

- Increased expectations from patients and medicolegal concerns.
- Increasing competition among surgeons.
- The availability of improved IOLs for treating presbyopia.
- The IOLMaster has raised the bar in AL accuracy.

Today the IOLMaster sets the standard for the accuracy and precision of AL measurements needed for IOL power calculations. The Partial Coherence Interferometer (PCI), however, is unable to obtain a reading in 8% to 20% of the patients due to dense cataracts (especially those with sub-capsular plaques) and fixation problems (especially in eyes with macular degeneration).

There are two main features that make the immersion method superior to applanation:
- No uncontrollable and uncorrectable applanation which shortens the eye on the average approximately 0.2 mm (shortening up to 0.8 mm has been observed).
- Objective alignment of the ultrasonic beam with the optical axis of the eye.

Basic Principles of the Immersion Method

- Ultrasonic probe remains remote from the cornea (no touch, no applanation) (Fig. 6-1).
- Display of both corneal signals together with both lens signals and the retinal (+ sclera) signals.
- Alignment of the ultrasonic beam with the optical axis of the eye (objective alignment much more reliable than subjective alignment).
- Optimal results are comparable to those obtained with the IOLMaster (requires, however, adequate instrument parameters and optimal techniques).

Examination Technique: Preparation

- Patient in semi-supine position.
- Local anesthetic.
- Flawless immersion (no leaking):
  - Immersion shells (ultra-light, adequate sizes, no need to be hand-held) (Fig. 6-2).
  - Saline as immersion fluid (no air bubbles in spite of repeated "in and out" of the probe) (Fig. 6-3).
  - Prior sealing of the inner shell edges with thicker fluid (eg, 1.6 % methyl-cellulose).

Examination Technique

- Hand-held probe immersed and grossly placed over corneal center. Observation needs to occur from above (through the fluid) and not from the side so as to avoid misleading refraction.

A-Scan Biometry: Avoiding Pitfalls With Immersion Technique

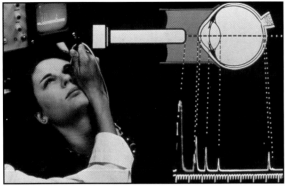

Figure 6-1. Immersion A-scan remains remote from the cornea preventing eye compression.

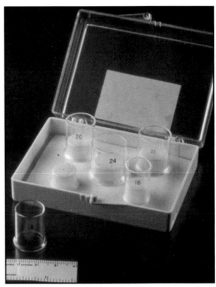

Figure 6-2. Hansen Ossoinig immersion A-scan shells.

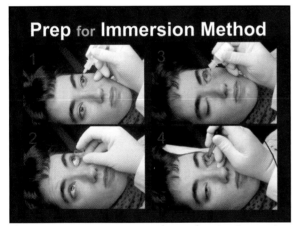

Figure 6-3. Preparation steps for performing immersion A-scan.

- Initially, put the A-scan on high instrument gain (high ocular signals from the cornea, both lens surfaces, and the fundus) to more easily recognize a suitable beam direction.
- Alignment of the ultrasonic beam with the axis of the eye at decreasing gain settings. There are two ways to achieve alignment:
  - *Subjective:* This is the traditional method using a light source in the tip of the US probe so the patient can fixate on the light. This approach is prone to error because the acoustic and optical pathways are not identical and may deviate from each other significantly. This can be due to different refractions, especially in lenses with dense subcapsular plaques. These are one of the sources of failure with the IOLMaster. Other patients unable to benefit from the IOLMaster are those not able to fixate on the light source at all.
  - *Objective:* This is optimal because the beam is objectively aligned with the optical axis of the eye. This approach secures optimal measuring accuracy. This is achieved when the ultrasonic beam is aimed perpendicularly at the cornea and both lens surfaces simultaneously. The retina signal is not assessed at this point.

This simultaneous perpendicularity automatically guides the beam toward the macula, and this is especially important in long eyes with posterior staphylomas and in short eyes with atypical shapes.

This objective alignment of the ultrasonic beam with the optical axis of the eye does not require patient cooperation. To obtain this alignment the probe must be freely movable by the hand of the examiner (do not use probe holders), which allows for such an objective beam alignment. The instrument gain is slowly decreased during the scanning procedure in order to reduce the height of the cornea and lens spikes as much as possible (without losing the presumed retinal signal). At low spike height, their status of being maximized (perpendicular sound beam) is recognized much better as well as adjusted (increased display sensitivity).

This optimal alignment of the ultrasonic beam with the optical axis of the eye (ie, aiming the beam from the corneal center to the fovea and thus measure the correct AL) leads to optimized "measuring accuracy." Is this difficult? Yes it is, but only in the beginning. With some practice an acquired reflex sets in which allows the hand of the examiner to guide the beam almost automatically into the needed simultaneously perpendicular direction according to the behavior of the signal height noted by the examiner while observing the screen display. This acquired reflex is comparable to how we are guided while walking, automatically activating and controlling the relevant muscles.

- Freeze the optimized (regarding beam position and direction) echogram, eg, by depressing a foot switch.
- Electronic measuring gates are usually set automatically once the echographer accepts an AL echogram as optimal (eg, by depressing a foot switch). They need, however, to be checked for precise setting and, if necessary, must now be corrected before the echogram is processed for calculation of the measured AL.
- When possible, peak-to-peak measurements should be applied. The peaks of the echo spikes truly represent the underlying acoustic interfaces and are not affected by phase shifts stemming from differences in echo intensity between the measured signals (eg, between the anterior and posterior lens signals). While such phase shifts cause only minimal errors in the distance measurements, they should be avoided for the sake of optimal measuring precision. Most biometric instruments, however, do not yet allow this approach. It is then best to get the measuring markers as close as possible to the peaks of the relevant echo spikes. The worst scenario is to measure from signal base to signal base.
- Next, check on the retinal signal by increasing the instrument gain and thus all of the echo signals to confirm that the presumed retinal signal indeed represents the retinal surface and not the stronger reflective sclera surface.
- Finally, send this echogram for processing the calculation of the measured AL in mm.
- Repeat this entire procedure at least three times for optimal results by repeating all the above steps from the beginning.
- The measuring procedure is completed when three acceptable echograms have been obtained. Repeating the entire procedure each time independent from each other one avoids making the same alignment mistake each time, which may also result in close (but false) values.
- The final result is the average of the three measurements.

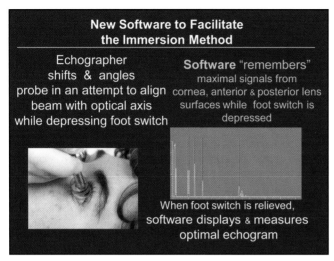

Figure 6-4. New software facilitates the use of the immersion method.

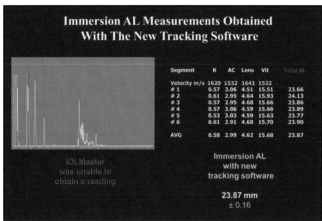

Figure 6-5. Example of results of new software facilitating the immersion method.

New Software to Improve the Immersion Method

An experimental design of a new software (developed by this author and Jean Abascal of Quantel Medical) was tested on a number of cataract patients in the University of Vienna Eye Clinic. The author performed the ultrasonic immersion measurements using the new software while Prof. Dr. Lukas measured the ALs of the same patients using the IOLMaster. Both measurements were then compared. In those cases where the IOLMaster was successful, the results were almost identical with the echographic measurements, with a tendency of the PCI measurements to be minimally longer.

With this software, all the echographer has to do is angle and shift the US beam, attempting to aim a perpendicular beam at both cornea and lens surfaces (Fig. 6-4). The instrument then "remembers" which were the optimal readings and, upon simultaneous display of maximize corneal and lens signals, freezes and measures this optimal echogram (beam precisely aligned with the optical axis of the eye) (Fig. 6-5).

Summary of Requirements for an Optimal Immersion Technique

- Immersion shells for different eye sizes.
- Hand-guided probe.
- Start at high system sensitivity for gross beam alignment.
- Decrease system sensitivity for selection of optimal beam alignment through simultaneous display of maximal signals from the corneal and lens surfaces. This objectively aligns the acoustic beam with the optical axis of the eye, while decreasing system sensitivity to optimal (low) setting.
- Optimal alignment determined by examiner vs. instrument software.
- Optimal echogram is frozen "manually" by the echographer or automatically by the new software.
- Five measuring gates using correct tissue-dependent sound velocities (cornea 1641 m/sec; anterior chamber and vitreous cavity 1532 m/sec; lens 1641 m/sec).
- If necessary, correct the automatically set optimal gate settings over the 5 peaks representing the corneal epithelium and endothelium, the anterior and posterior lens surfaces, and the retinal surface.
- Increase system sensitivity to verify retinal signal in frozen echogram.
- Obtain 3 independent measurements with close results repeating each time the entire procedure beginning with probe immersion and including optimal beam alignment.
- Use the average AL measured this way (calculated and displayed automatically by the instrument).

Instrument and Probe Designs for the Immersion Method

Because of its easier application (both facility- and technique-wise) the Applanation Method dominated the market worldwide. As a consequence, the technical requirements for biometric instruments were mostly designed and continuously improved for the Applanation Method with all its faults. The requirements for the Immersion Method were largely ignored by the majority of the instrument makers. Today that trend is in reverse, but most instrumentation in use is partly (some entirely) unfit to achieve full success using immersion. Therefore, at this time, many echographers will not yet be able to utilize all the advantages of the Immersion Method.

Mandatory Required Parameters

- Correct sound velocities (see Table 4-1).
- Adjustable instrument gain at least prior to freezing of echogram.
- Dynamic Range ("acoustic field") between 25 and 45 db.

Strongly Preferred Parameters (All Realized in Standardized Instruments)

- Use a non-focused 8 MHz probe with parallel beam to allow easy recognition of beam perpendicularity and application of the objective beam alignment with the optical axis of the eye. Higher frequencies and smaller probes all require focusing their beams and consequently cause difficulties in recognizing beam perpendicularity.
- S-shaped amplifier characteristics (an optimal compromise between linear and logarithmic amplifiers, each of which has critical advantages and shortcomings).
- Dynamic range ("acoustic field") of 36 dB.
- Observe spike configuration to assure optimal display (see Fig. 6-5).

Optimal Probe Properties

- A parallel (non-focused) ultrasonic beam provides more accurate and precise measuring results than a focused beam because signals rise sharply only when the ultrasonic beam reaches an acoustic surface perpendicularly. It thus allows the examiner to recognize perpendicularity while maximizing the signal height.
- Frequency of the transducer in the probe is of secondary importance. While it is true that higher frequencies provide better resolution, they must be focused to avoid divergent beams and so far they have never been focused to the point of emitting a parallel beam.
- In standardized instruments an 8 MHz standard probe is always used. Its transducer diameter of 5 mm provides an emitted parallel beam. By reducing the instrument gain, this parallel beam can be narrowed as needed for optimal resolution and reducing the spike height brings the peaks into the range with greatest sensitivity (ie, low to medium spike height). Thus this clearly bigger probe provides the facility of recognizing beam perpendicularity while the smaller focused probes do not. By reducing the instrument gain, the beam can be made very narrow yet still achieve the advantages of a focused beam without losing the parallel feature.

Optimizing Results Using Non-Standardized A-Scans: Optimizing Measuring Precision

- Use parallel rather than focused ultrasonic beam (to be able to assess correct beam direction).
- Use logarithmic rather than linear amplification (to avoid false maximal appearance of signals).
- Start with high instrument sensitivity. Reduce instrument sensitivity to a minimum still displaying all pertinent signals before measurement.
- Maximize signals from cornea and lens simultaneously to optimize beam direction.

Figure 6-6. Differentiation between lens and iris signals when the pupil is not dilated.

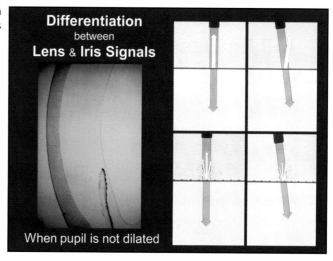

- Do peak-to-peak measurements[3] of relevant spikes or, if not possible, set the measuring gates on the left side of the peaks as close to them as possible.
- Avoid devices for easier probe holding which are likely to hamper free probe movement and optimal beam alignment.

Avoiding Pitfalls of the Immersion Method

- When using a focused probe (made for the applanation method) removed from the corneal surface, the retinal surface may shift into the more remote divergent part of the beam. To correct this, acquire a probe from the manufacturer made for the immersion method. The best is a standard 8 MHz non-focused probe emitting a parallel ultrasonic beam.
- When the pupil is not dilated, the anterior lens surface signal may be confused with the iris signal which is not necessarily received first. Also, in addition, a lenticular subcapsular plaque signal may appear, further confusing the situation. To correct this, dilate the pupil before doing an immersion technique. If this is not possible (Fig. 6-6), try to differentiate between the signal from the iris surface (which is extremely coarse) and the signal from the anterior lens surface (which is extremely smooth). This is relatively easy when using a standardized instrument. Slight angling of the beam reveals each of these two competing signals. The lens signal readily disappears whereas the iris signal persists during the angling. When using a focused ultrasonic beam this differentiation may become quite difficult or even impossible.

References

1. Janson F, Koch E. Determination of the velocity of ultrasound in the human lens and vitreous. *Acta Ophthalmol (Kbh)*. 1963;40:420-433.
2. Janson F. Measurements of intraocular distances by ultrasound. *Acta Ophthalmol (Kbh)*. 1963;74(Suppl):1-51.
3. Ossoinig KC. *Basics of Standardized Ophthalmic Echography.* 2009 (CD); available from Echography Teaching Services at www.echography.com.

Available A-Scan Instrumentation

Kenneth J. Hoffer, MD

It is the purpose of this chapter to provide information and specifications for a variety of A-scans available and in use throughout the world. They are shown in Table 7-1. Some of these instruments listed in the table may no longer be available from the original manufacturer, but there is an international secondary market of used and refurbished equipment where those instruments may be an option.

Since the introduction of the Zeiss IOLMaster in 1999, interest in ultrasound A-scans for measuring AL has declined. However, A-scan examination is still necessary in 10% to 17% of eyes (depending upon the patient population) because the IOLMaster cannot obtain an accurate AL reading due to dense opacity, dense PSC cataract, or inability to fixate. Also the expense of the IOLMaster precludes its use in many parts of the world.

Table 7-1.

ULTRASOUND INSTRUMENT SPECIFICATIONS FOR A VARIETY OF A-SCANS FROM AROUND THE WORLD

Manufacturer	Instrument	Type	Software Formulas	Features
Accutome	AccuSonic	A-scan	Hoffer Q, Holladay, SRK/T	Large 7.5 inch, VGA resolution. LCD screen with intuitive user interface. Rotary user input knob and dedicated tactile buttons. Compact, portable unit. Optional keyboard, footswitch and external printer.
Accutome	Advent A/B System	Combo	Hoffer Q, Holladay, SRK, SRK/T	*A-scan:* Tonometer mounted, hand-held and immersion techniques. Multi-tone A-scan lock-on signal. Customized surgeon lens profiles. Calculation of 3 IOLs at once. Defaulted velocity settings. Patient archive data. *B-scan:* Pan and zoom feature. Window mode, update mode. 3 defaulted range values. Memory storage/image download.
Alcon	Ocuscan RxP	A-scan	Hoffer Q, Holladay, SRK/T	*NOT AVAILABLE IN USA* Biometry/pachymetry: desktop, computer interfacec, ontact or immersion, velocity programmable for each segment, personalization, and allows up to 5 users each.

(continued)

Table 7-1 continued.

ULTRASOUND INSTRUMENT SPECIFICATIONS FOR A VARIETY OF A-SCANS FROM AROUND THE WORLD

Manufacturer	Instrument	Type	Software Formulas	Features
Alcon	UltraScan	Combo	Hoffer Q, Holladay, SRK/T	*NOT AVAILABLE IN USA* B-scan, A-scan Biometry plus unique B-Biometry. Multifunctional system. Standard 10 MHz probe, or optional 20 MHz probe that brings 100 microns of resolution (depths of 30 to 35 mm) to posterior segment images.
DGH Technology	DGH 5000e	A-scan	Binkhorst II, Holladay, SRK II, SRK/T	Solid tip probe, 3 mm diameter, fully automatic, internal printer, real time graphic display. Pachymeter mode (5100e) simultaneously shows corneal thickness, current measurement position, and the selected corneal map. Multiple doctor's configurations. Data storage. Automatic measurement mode. No foot pedal is required. Exclusive corneal compression detection software.

(continued)

Table 7-1 continued.

ULTRASOUND INSTRUMENT SPECIFICATIONS FOR A VARIETY OF A-SCANS FROM AROUND THE WORLD

Manufacturer	Instrument	Type	Software Formulas	Features
DGH Technology	DGH 5100e	A-scan	Binkhorst II, Holladay, SRK II, SRK/T	Solid tip probe, 3 mm diameter, fully automatic, internal printer, real time graphic display. Pachymeter mode (5100e) simultaneously shows corneal thickness, current measurement, position and the selected corneal map. Multiple doctor's configurations. Data storage. Automatic measurement mode. No foot pedal is required. Exclusive corneal compression detection software.
Innovative Imaging Inc	I³ System-ABDv1	Combo	Hoffer Q, Holladay	*A-scan:* Measurement accuracy: 0.05 mm, store up to 20 A-scans internally. *AL Biometry:* Pre-programmed velocities, manual/auto modes, immersion or contact settings (ABDv2), solid 10 MHz probe.
Innovative Imaging Inc	I³ System-ABDv2	Combo	Hoffer Q, Holladay, SRK/T	*Standardized diagnostic:* Two gate measurements display, tissue sensitivity value stored, 8 MHz parallel beam probe. *B-scan:* True geometry with stepped zoom, pan, freeze frame, post image processing. Posterior segment 10 MHz probe. Wide-field anterior segment 20 MHz probe.

(continued)

Table 7-1 continued.

ULTRASOUND INSTRUMENT SPECIFICATIONS FOR A VARIETY OF A-SCANS FROM AROUND THE WORLD

Manufacturer	Instrument	Type	Software Formulas	Features
Nidek Inc	Echoscan US 800	A-scan	Binkhorst II, Holladay, SRK II, SRK/T	Gate function; 40 mm measure range; portable, compact, and lightweight. Speedy and highly accurate measurement. Five IOL power calculations.
Nidek Inc	Echoscan US 1800	A-scan	Binkhorst, Hoffer Q, Holladay, SRK, SRK II, SRK/T	Adjustable amplifier gain. Measuring range 12 to 40 mm. Ten MHz solid with internal red fixation axial length measurement probe.
Nidek Inc	Echoscan US 2520	Combo	Binkhorst II, Holladay, SRK II	Interfaces with US 2500 for A/B combination.
Nidek Inc	Echoscan US 3300	Combo	Binkhorst II, Holladay, SRK II	Versatile A/B-scan. Post processing function. User-friendly operation. Biometry probe is optional.
Ocuserv	DB-3000		Hoffer Q, Holladay, SRK II, SRK/T	Ten MHz focused transducer with fixation light. Stores 5 scans per eye, built-in printer. Can be upgraded to B-scan.
Ocuserv	DB-3000C		Hoffer Q, Holladay, SRK II, SRK/T	Light and affordable. Ten MHz focused transducer with fixation light. Stores 5 scans per eye, printer with results of ACD and lens thickness.

(continued)

Table 7-1 continued.

ULTRASOUND INSTRUMENT SPECIFICATIONS FOR A VARIETY OF A-SCANS FROM AROUND THE WORLD

Manufacturer	Instrument	Type	Software Formulas	Features
Ocuserv	DB-3000CG		Hoffer Q, Holladay, SRK II, SRK/T	Light and affordable. Ten MHz focused transducer with fixation light. Stores 5 scans per eye, printer with results of ACD and lens thickness.
Ocuserv	DB-3100		Hoffer Q, Holladay, SRK II, SRK/T	Ultrasound software on a laptop computer with probe attached. Ten MHz focused transducer. Uses computer printer.
Ophthalmic Technologies Inc	OTI-A2000		Hoffer Q, Holladay, SRK II, SRK/T	Compact, portable 13 Mhz A-scan connects to any laptop or PC computer. Easy-to-use software automatically selects and ranks scans. Up to 9 scans can be displayed side by side for fast comparison. Advanced algorithms for high myopia, pseudophakia, and dense cataracts. All major IOL formulas and unlimited IOLs.
Ophthalmic Technologies Inc	OTI-2000	Combo	Hoffer Q, Holladay, SRK II, SRK/T	Ten or 20 MHz unfocused transducer A-scan connects to any laptop or PC computer. Easy-to-use software automatically selects and ranks scans. Up to 9 scans can be displayed side by side for fast comparison. *Dynamic Digital Recording* B-scan. Advanced algorithms for high myopia, pseudophakia and dense cataracts.

(continued)

Table 7-1 continued.

ULTRASOUND INSTRUMENT SPECIFICATIONS FOR A VARIETY OF A-SCANS FROM AROUND THE WORLD

Manufacturer	Instrument	Type	Software Formulas	Features
Ophthalmic Technologies Inc	OTI-B/A /3D 1000	Combo	Hoffer Q, Holladay, SRK II, SRK/T	Real time dynamic movie recording, Integrated 3D, 12 MHz B-scan, 13 MHz A-scan, UBM option, 50 frames/sec, portable and compact. Desktop or laptop models available.
Optikon	Bioline	A-scan	Binkhorst, Haigis, Holladay, SRK II, SRK/T	Compact, transportable unit with built-in printer. Large screen LCD blue display. Waterproof keyboard. Immersion biometry. Stores 15 IOLs.
Optikon	Hi-line	A-scan B-scan, UBM	Binkhorst, Haigis, Holladay, SRK II, SRK/T	Biometric A-scan is provided through 10 MHz biometry probe. The biometry probe can be hand-held, mounted on a tonometer, or provided with accessories for immersion. The software ensures differential measurements for pseudophakic eyes in relation to various lenses.
Paradigm Medical Industries Inc	P20	A-scan	Binkhorst, Hoffer Q, Holladay, SRK II, SRK/T	Maintenance-free, solid tip, lighted probe. Automatic, semi-automatic and manual mode operations. Hands-free operation. Complete display of scan and intraocular dimensions, with select screening and editing. Quiet, built-in printer.

(continued)

Table 7-1 continued.

ULTRASOUND INSTRUMENT SPECIFICATIONS FOR A VARIETY OF A-SCANS FROM AROUND THE WORLD

Manufacturer	Instrument	Type	Software Formulas	Features
Paradigm Medical Industries Inc	A/B P37	Combo	Binkhorst, Hoffer Q, Holladay, SRK II, SRK/T	Customized and user-friendly. Superior Quad Imaging. Built-in flexibility. Standard biometry and diagnostic echography. Precision A and B probe.
Quantel Medical	Aviso B	Combo	Hoffer Q, Holladay, SRK II, SRK/T	AVISO remote control touch screen, 24" Dell desktop or laptop (pre-configured with AVISO software), 10 MHz B-Scan probe, foot switch, power strip, keyboard, and printer.
Quantel Medical	Axis II	A-scan	Hoffer Q, Holladay, SRK II, SRK/T	Carrying case, biometry probe with extension handle, foot-switch, Prager immersion shell, printer with cable, power strip.
Quantel Medical	Compact 2B	Combo	Hoffer Q, Holladay, SRK II, SRK/T	Ten MHz B Probe, power strip, foot switch, integrated keyboard/trackball/mouse, 9" cube monitor and video printer.
Quantel Medical	Cinescan B or S	Combo	Hoffer Q, Holladay, SRK II, SRK/T	Built-in monitor, 10 MHz B probe, foot switch, integrated keyboard/trackball/mouse, video printer. Precision biometry software, probe, extension handle (for applanation), Prager immersion shell, printer S = Standardized A-scan by Karl C. Ossoinig, MD.

(continued)

Table 7-1 continued.

ULTRASOUND INSTRUMENT SPECIFICATIONS FOR A VARIETY OF A-SCANS FROM AROUND THE WORLD

Manufacturer	Instrument	Type	Software Formulas	Features
Sonogage	EyeScan	A-scan	Holladay, SRK II, SRK/T	Solid tip probe 10 MHz transducer, automatic pattern recognition with tonal alert, internal dot matrix printer. Accuracy ±0.034 mm claimed.
Sonomed Inc	Microscan 100 A+	A-scan	Binkhorst, Holladay, Hoffer Q, SRK II, SRK/T	Live A-scan display. Five different examination modes. Measure review capability. Clinical accuracy ±0.1 mm. Optional printer. Two probe styles available.
Sonomed Inc	PacScan 300AP	A-scan	Binkhorst, Hoffer Q, Holladay, SRK II, SRK/T	Touch screen operation. Large, high-resolution, backlit LCD. Live A-scans display. Storage of 5 different user profiles. Five different examination modes. Measurement review capability. Immersion capabilities. Clinical accuracy ±0.1 mm. Optional printer.
Sonomed Inc	A5500	A-scan	Binkhorst, Hoffer Q, Holladay, SRK II, SRK/T	Three different probes. Four different modes. Personalized A-constants and surgeon factors. Clinical accuracy ±0.1 mm. Built-in calibration cylinder.
Sonomed Inc	A/B5500	Combo	Binkhorst, Hoffer Q, Holladay, SRK II, SRK/T	Incorporates all of the features of the A and B units with advantage of compactness.

(continued)

Table 7-1 continued.

ULTRASOUND INSTRUMENT SPECIFICATIONS FOR A VARIETY OF A-SCANS FROM AROUND THE WORLD

Manufacturer	Instrument	Type	Software Formulas	Features
Sonomed Inc		Combo	Binkhorst, Hoffer Q, Holladay, SRK II, SRK/T,	Incorporates all of the features of the A and B units with advantage of being housed in a laptop computer.
Storz Ophthalmics	CompuScan LT Biometric	A-scan	Binkhorst II, Holladay, SRK II, SRK/T	Company was purchased by B+L and no longer sells A-scans.
Storz Ophthalmics	CompuScan AB	Combo	Binkhorst, Hoffer Q, Holladay, SRK/T	Company was purchased by B+L and no longer sells A-scans.
Tomey Corporation	AL-100	A-scan	Haigis, Haigis optimized Hoffer Q, Holladay, SRK II, SRK/T	Measures axial length and calculates IOL power. Easy to use touch screen. Automatic tone-assisted measurement. Compact and lightweight. Built-in printer. Optional memory card. Contact or immersion modes.
Tomey Corporation	AL 2000	A-scan	Haigis, Haigis optimized Hoffer Q, Holladay, SRK II, SRK/T	Measures axial length, corneal thickness and calculates IOL power. Ultrasound technology provides precise measurements. Easy to use touch screen maximizes efficiency. Compact and lightweight. Contact and immersion mode. Multipoint pachymeter map. Wide measurement range. Solid state probes.

(continued)

Available A-Scan Instrumentation 49

Table 7-1 continued.

ULTRASOUND INSTRUMENT SPECIFICATIONS FOR A VARIETY OF A-SCANS FROM AROUND THE WORLD

Manufacturer	Instrument	Type	Software Formulas	Features
Tomey Corporation	AL 3000	A-scan	Haigis, Hoffer Q, Holladay, SRK II, SRK/T	Measures axial length and calculates IOL power. Easy to use touch screen maximizes efficiency. Contact and immersion mode. Wide measurement range. Solid state probes.
Tomey Corporation	UD 1000 & 6000	Combo	Haigis, Haigis opt Hoffer Q, Holladay, SRK II, SRK/T	High resolution annular array B-scan probe. Automated video recording. Area and distance measurement. Touch screen. Memory card. Optional diagnostic A-scan. A-scan (UD 6000) measures axial length and calculates IOL power.
Wuxi Kangming Medical Device Co (China)	KN-1800	A-scan	SRK I, SRK II, SCDK, & KORA	Solid probe 10 MHz with focus light. Equipped with special hardware and software to compose Callan myopia forecasting and forewarning system.

Some of the above instruments may no longer be available for new purchase but may be available on the secondary/used/refurbished market.
The reference to the Holladay formula means the Holladay 1 formula; no instrument has the Holladay 2 formula installed at this time.

(continued)

Table 7-1 continued.

ULTRASOUND INSTRUMENT SPECIFICATIONS FOR A VARIETY OF A-SCANS FROM AROUND THE WORLD

Optical (Non-US) Instruments for AL Measurement

Manufacturer	Instrument	Type	Software Formulas	Features
Zeiss Meditec Jena, Germany 1999	IOL Master	Optical	Haigis, Haigis L, Hoffer Q, Holladay, SRK II, SRK/T	Advantages of laser optical biometry in a non-contact technique: Measuring precision and repeatability are high. It is fast and easy for the patient. Measures AL, K, ACD, CD.
Haag-Sreit Koeniz, Switzerland 2009	LENSTAR LS 900	Optical	Hoffer Q, Holladay, SRK II, SRK/T	Most recent published studies demonstrate equivalence of the LENSTAR to the IOLMaster in accuracy and repeatability. It is not as fast, but in one pass it measures: AL, K, CT, AQD, LT, CD, RT.

Figure 7-1. Immersion method; probe dipped into Ossoinig shell.

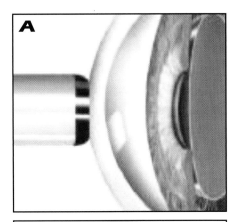

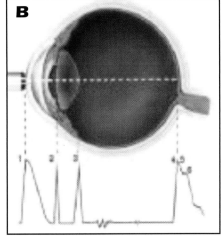

Figure 7-2. Contact or applanation method.

The authors of this book emphasize the importance of using the immersion method (Fig. 7-1) rather than the contact or applanation methods (Fig. 7-2). That is the reason there is no chapter on applanation in this book. It is important to know which techniques the instrument you are considering can perform, otherwise it may not enable you to step up to the immersion method.

One of the options with A-scans is the type of probe that is used and whether the transducer is focused or non-focused. It appears to make sense that focusing the sound wave at a distance of 23 to 24 mm to reach the retina would be a good idea. However, Ossoinig strongly recommends a nonfocused transducer because the sound wave may be focused for a 23.5 mm eye but the wave will be a converging wave in a shorter eye and a diverging one in a longer eye. This will create a greater error than is seen with a nonfocused transducer.

Note: Information on specifications were obtained from respective manufacturers.

Immersion Using the Prager Shell

Thomas C. Prager, PhD, MPH

Applanation Biometry

Contact, or applanation biometry, begins with the ultrasound probe being placed directly on the surface of the cornea. In contrast, the immersion technique utilizes a liquid interface between the eye and the ultrasound probe. While direct applanation on the cornea seems simple enough and is probably used more often than the immersion technique, contact biometry is the least accurate method of measuring axial length. When trying to place the probe tip on the center of the cornea, even experienced technicians may encounter parallax problems, resulting in a measurement that is off axis by small or large amounts. Remember that the eye is the longest when measuring through the center of the cornea. Secondly, when attempting to center the probe directly on the cornea, there will *always* be some degree of corneal compression, especially in patients with reduced intraocular pressure. Short eye measurements as a result of corneal compression associated with the applanation technique will readily reduce the accuracy required for presbyopia-correcting IOLs.

The Immersion Technique Using a Fixed Immersion Shell

The Prager Shell was designed in 1982, and represented an improvement upon an immersion shell first created by Jackson Coleman, MD.[1] The immersion technique, a one-handed procedure that is easy to master, eliminates or minimizes technician variables (such as corneal compression, alignment of the ultrasound beam, and probe insertion depth) and leads to more reproducible results. Each biometer manufacturer has specified an optimal distance from the cornea for data acquisition; having an immersion shell with a fixed shelf or a probe auto-stop ensures reliable probe placement. The fixed immersion technique is much easier to master than the open shell technique, which requires a certain level of dexterity in order to position the probe at the appropriate distance from the cornea while simultaneously being perpendicular to the retina and directing sound waves through the center of the cornea and lens. Immersion biometry utilizing the Prager Shell has been reported to be faster than the contact method.[2] Aside from ease in getting on axis, time is saved by not having to review individual scans for corneal compression errors.

In the quest of greater accuracy in surgical outcome, there have been many comparisons between applanation and immersion techniques first investigated by Hoffer in 1981 and published with Shammas[3] in 1984, who reported that immersion scans consistently result in longer axial lengths and less variability than the contact technique. This has been replicated in many studies[2-10] over the past 30 years. The Prager Shell and other immersion devices have been directly compared to the IOLMaster (non-contact coherent light method of AL determination) and no clinical measurement differences have been found between the two methodologies, although there is a significant difference in the cost of the equipment.[2,5,11] Further, because the IOLMaster was specifically set to match immersion and not applanation, there is a high correlation between immersion biometry and the IOLMaster units in AL measurements (Pearson correlation coefficient = 0.996).[2,12] AL measurements using the IOLMaster are unobtainable in 8% to 17% of the cataract population due to reduced visual acuity, corneal and media opacities, as well as dense cataracts.[11-16]

Thus, immersion biometry will always have clinical utility.

Consideration When Using the Fixed Immersion Shell

The balance of this chapter will discuss practical hints and pearls when using the fixed immersion Prager Shell and suggestions on reducing measurement error. It is important to emphasize that just utilizing an immersion shell does not guarantee perfect results in every patient. There is no substitute for understanding the sources of error when performing immersion biometry. One must be able to recognize an unusual scan result and be interested and compulsive enough to resolve any apparent discrepancy. I often say, "An inexperienced technician, who does not understand the basic principles underlying AL scanning, can undermine the efforts of the most skillful cataract surgeon."

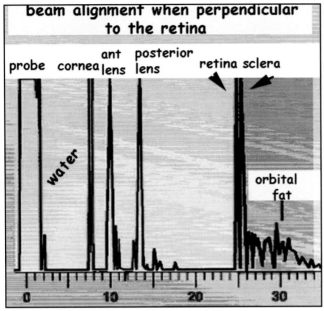

Figure 8-1. When perpendicular to the retina, A scanspikes are of equal height and both retinal and scleral heights are within 80% of one another with no stair stepping of the retinal spike.

Note that all spikes are of equal height and both retinal and scleral heights are within 80% of one another with no stair stepping of the retinal spike (Fig. 8-1). If one does not see spikes associated with orbital fat, the scan may be going through the optic nerve and not the fovea. A B-scan is required when encountering a wide separation between retinal and scleral spikes which could indicate edema. Any eye which does not have a clear cornea and/or vitreal media will require a sonogram.

REVIEW THE CHART PRIOR TO MEASURING THE EYE

Most refractive surprises occur in unusual length eyes and are more prevalent in short eyes. Proportionately a 1 mm error in a 20 mm eye has a greater postoperative refractive consequence than the same 1 mm error in a 30 mm eye. Although staphylomas may make it more difficult to locate foveal spikes in a long eye, the short eye requires greater measurement precision. Prior to an ocular measurement it is the responsibility of the biometrist to determine if the eye is unusual in any way. Peruse the chart because a previous scleral buckle can change the shape of the eye producing a significant AL difference between the eyes. For almost all patients both eyes are approximately the same length, typically within 0.3 mm of one another. Replicate measurements several times and if a 0.3 mm or greater difference remains, note in the chart that the "measurements exceed normal physiological findings." To verify this difference in axial length, obtain a B-scan through the optic nerve of both eyes which will allow a direct comparison. Display the two scans (with x-axis gradations) one above another, then draw a line from the optic nerve of one sonogram down to the x-axis and continue to the same axis location on the other echogram (Fig. 8-2). Subtle differences are readily seen and this documentation should be included in the patient chart.

Determine if either eye is aphakic, as this will require a change in sound velocity to compensate for the missing lens. Similarly, a pseudophakic eye will require a change in tissue velocity. Most biometers have settings for the different types of IOL materials which must be known for an accurate axial length (see Chapter 4).

Figure 8-2. Comparing the B-scan through the optic nerve of both eyes to verify a reported difference in AL.

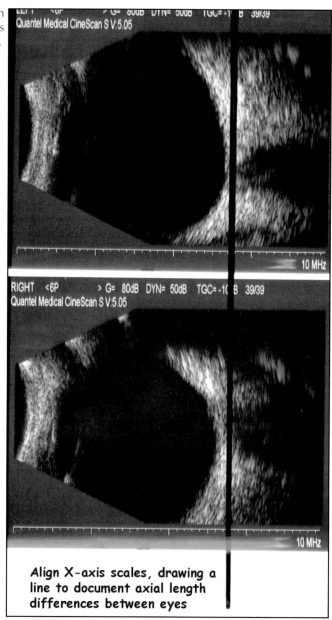

SILICONE OIL EYES

Measuring eyes that contain silicone oil in the posterior vitreous is a difficult task with immersion biometry and, in this author's experience, seldom results in an accurate AL measurement. Silicone oil in the eye slows the speed of sound through the eye, resulting in an artifactual long eye measurement. A further complication is that the two most common types of silicone each have different tissue velocities, 1050 or 980 m/s, and the clinician must discern which velocity to select to prevent a measurement mistake. Sandra Frazier Byrne addresses the problem of silicone[16] in AL measurements and outlines a

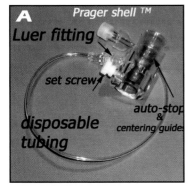
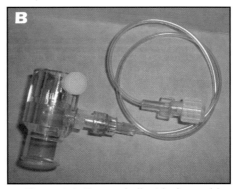
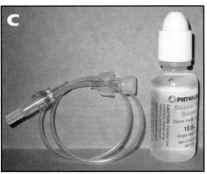

Figure 8-3. Prager shell with disposable tubing.

corrective procedure. Each ocular component of the eye, cornea, anterior chamber, lens, and posterior vitreous cavity must be measured individually and then summed. While the anterior chamber does not require a mathematical adjustment, to obtain true values for the lens, multiply X (the measured lens thickness) by 1641/1532 and multiply the measured vitreous length by 980 (most cases)/1532. Final IOL power determination is made more confusing because the index of refraction differs in the eye with silicone versus the normal intact eye and requires the addition of more refractive power. In the silicone oil eye, the IOLMaster produces the more accurate measurement.

Carefully review the chart for additional anomalies that can potentially affect the surgical outcome. What is the IOL power requested? If there is an anticipated 2 diopters difference or more in the final refraction and the other eye does not require cataract surgery, patients may not be able to tolerate the anisometropia. Prior to a measurement use the current glasses prescription to estimate the anticipated axial length. The average eye is 23.5 mm, so given that 1 mm = ~2.35 D, a 24-mm eye should be roughly myopic by ~1.2 D. If the glasses have a hyperopic optical correction of +3.00 diopters and the AL measurement is 26 mm, this should alert the biometrist that there is a potential mistake.

PRACTICAL TIPS IN OBTAINING IMMERSION ANTERIOR-POSTERIOR LENGTH MEASUREMENTS

The Centers for Disease Control (CDC) guidelines require thoroughly soaking the shell and probe in a beaker of alcohol or hydrogen peroxide for at least 5 minutes prior to measuring the eye.[16] Subsequently, allow the immersion shell to completely dry or flush with BSS, since alcohol can readily remove corneal epithelial cells (Fig. 8-3). To reduce the likelihood of transmitting pathogens from patient to patient, change the connecting

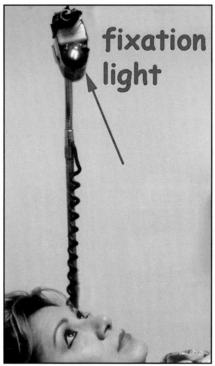

Figure 8-4. Fixation light on extension pole.

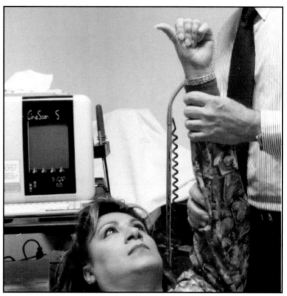

Figure 8-5. Examination of one-eyed patient.

tubing with each patient. Tubing with a check valve prevents reflux and will permit the use of the same bottle of BSS over several patients. A study[17] conducted at 34 ophthalmology clinics showed wide variability in shell/probe/tubing cleanliness, with only 14% of the centers following CDC guidelines. Fungus was cultured in 12% of the samples and microorganisms associated with endophthalmitis or keratitis were found in 53% of the sites tested. The Prager Shell has a Luer fitting to facilitate changing the tubing.

If there is an outer plastic sheath for applanation, pull it away from the probe. Since the acquisition software will capture scans *only* if the probe (main bang) is at the specified distance from the cornea specified by each ultrasound manufacturer, insert the probe into the Prager Shell until it seats at the auto-stop, then gently tighten the external set screw. Internal centering guides hold the probe in place to ensure probe perpendicularity within the shell. Routine topical anesthesia is administered to each eye. Seat the patient with their head tilted slightly back against a counter or use a reclining examination chair. A fixation light on a flexible stem facilitates this biometry (Fig. 8-4) but make sure that it is far enough from the patient's eyes to avoid convergence, which increases the difficulty of locating the fovea during biometry. The patient's attention is naturally drawn to the fixation light, which is beneficial if you do not speak their language.

When examining the one-eyed patient, proper fixation can be a problem. A helpful hint is to have the patient extend their arm, make a fist and have them stare at their thumb (Fig. 8-5). Even if the patient is completely blind, through proprioceptive feedback the eye will be able to locate and follow. Always support and move the arm to minimize fatigue.

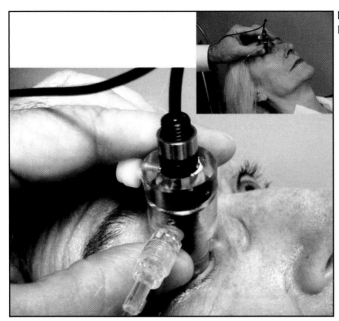

Figure 8-6. Proper placement of the Prager shell between the lids.

Examining Technique

Place a towel on the patient's shoulder; secure the tubing and BSS bottle and syringe to the shell Luer fitting. Ask the patient to look downward, toward their feet; then lift the patient's upper eye lid and insert the flared rim underneath the lid (the upper portion of the shell will make contact with the sclera while the lower part of the shell will be held away from the eye); ask the patient to look straight ahead with the uncovered eye, toward the fixation light. Pull the patient's lower eyelid down and gently pivot the lower portion of the shell into the lower fornix, making sure by close inspection that it is in the fornix and *not* sitting atop a fold in the conjunctiva. This pivotal motion avoids contact with the cornea and insures centration of the device around the limbus.

How to Hold the Shell

The goal is to put minimal pressure on the eye. In fact, it is quite instructive for *YOU* to be the patient (at least once) and experience firsthand the benefits of a light touch.

Note the Luer filler port is faced temporally (Fig. 8-6). The left hand/palm is resting on the forehead (given the biometry instrument is to your left), and is used to reduce shell pressure on the eye. Try to keep the A-scan instrument in your direct line of sight. It is important to position the biometer screen so that it can be seen easily during the procedure. Looking over your shoulder would needlessly complicate the examination. Note that the palm acts as the fulcrum or pivot point for the shell. The examiner may want to stabilize the shell with the right hand to make micro-movements. With practice, most practitioners usually will hold the shell with just the hand resting on the forehead. The right hand is free to make instrument adjustments, if necessary.

To make a measurement, slowly inject the saline into the shell. Quickly jetting fluid into the shell may remove corneal epithelial cells, resulting in patient discomfort. As soon as the liquid fills the shell sufficiently to reach the tip of the probe (about 2 cc), the typical waveforms of immersion biometry, as previously illustrated, will be seen on the screen. Most commercial biometers capture at least 10 scans and display an average and a standard deviation. The standard deviation, a measure of variability, should be less than 0.05. If this value is larger, review the individual waveforms and delete the outliers. Do not remove the shell from the patient's eye during this review, as any measurement that is deleted will need to be immediately replaced with a new and acceptable reading.

To remove the shell from the eye, raise the patient's upper eyelid, which releases the top part of the shell from under the eyelid. Next, pivot the shell downward, directing the patient to continue to look straight ahead. Then pull away from the eye without contacting the cornea. Upon the initial release, the remaining contents of the shell (1 to 2 cc of liquid) will spill down the patient's cheek. Be prepared with a towel or facial tissue.

CRITICAL BIOMETRY TIPS

Although the Prager Shell completely eliminates corneal compression as a complicating factor, and greatly assists in the alignment of the probe with the macula, it is still necessary to review and analyze waveforms to insure optimum readings.

Again, be sure to accept only steeply rising retinal spikes and reject any that have a stair-step appearance. The corneal, anterior lens, and retinal spikes should be of approximately equal height. If the spikes demonstrate a downward trend, this suggests that the scan is off-axis. With dense cataracts, the tendency is to increase the gain thereby elevating the spikes. If the tops of the spikes appear flattened, this may indicate that the amplifiers are saturated, resulting in an inaccurate reading.

With very long eyes, such as in those with staphyloma, the macula may be located on the sloping portion of the staphyloma and the retinal spike may not rise to the same height as the corneal spikes. However, in normal eyes, the retinal/scleral spikes equal the height of the corneal spikes. Detection of orbital fat spikes is a requirement. A normal scan has a series of orbital fat echoes with descending amplitudes. If they are absent, or markedly attenuated, the probe may be misaligned and the biometrist may have directed the sound beam to the optic nerve instead of the fovea.

Summary of Critical Points

- Always measure both eyes in every patient.
- Re-measure both eyes if:
    - There is >0.03 mm difference between eyes.
    - AL is <22 mm or >25 mm in either eye.
    - AL does not correlate well with patient's spectacle refraction.
    - There is difficulty obtaining correctly positioned, high, steeply rising echoes that are not stair-stepped. The retina and scleral spikes are within an 80% height of one another.
    - There are problems with patient cooperation or fixation.
    - Perform a B-scan to document the difference.
- When doing the second eye later, always measure the first eye again and make a post-cataract comparison to adjust the surgeon factor in the IOL formula.

Once the biometrist is familiar with the basics discussed in this chapter, the learning curve is probably 5 to 10 patients and you will have mastered the Prager shell immersion technique and have confidence in your measurements.

References

1. Coleman DJ, Silverman RH, Lizzi FL, et al. *Ultrasonography of the Eye and Orbit,* 2nd ed. Philadelphia: Lippincott Williams & Wilkins; 2005.
2. Packer M, Fine IH, Hoffman RS, Coffman PG, Brown LK. Immersion A-scan compared with partial coherence interferometry: Outcomes analysis. *J Cataract Refract Surg.* 2002;28(2):239–242.
3. Shammas HJ. A comparison of immersion and contact techniques for axial length measurements. *J Am Intraocul Implant Soc.* 1984;10(4):444–447.
4. Fries U, Hoffmann PC, Hut WW, Echardt HB, Heuring A. IOL-calculations and ultrasonic biometry: Immersion and contact procedures. *Klin Monatsbl Augenheilkd.* 1998;213:162–165. German. No abstract available.
5. Giers U, Epple C. Comparison of A-scan device accuracy. *J Cataract Refract Surg.* 1990;16(2):235–242.
6. Hoffmann PC, Hutz WW, Eckhardt HB, Heuring AH. Intraocular lens calculation and ultrasound biometry: Immersion and contact procedures. *Klin Monatsbl Augenheilkd.* 1998;213(3):161–165. German.
7. Olsen T, Nielsen PJ. Immersion versus contact technique in the measurement of axial length by ultrasound. *Acta Ophthalmol* (Copenh). 1989;67(1):101–102.
8. Schelenz J, Kammann J. Comparison of contact and immersion techniques for axial length measurement and implant power calculation. *J Cataract Refract Surg.* 1989;15(4):425–428.
9. Watson A, Armstrong R. Contact or immersion technique for axial length measurement? *Aust N Z J Ophthalmol.* 1999;27(1):49–51.
10. Haigis W, Lege B, Miller N, Schneider B. Comparison of immersion ultrasound biometry and partial coherence interferometry for intraocular lens calculation according to Haigis. *Graefes Arch Clin Exp Ophthalmol.* 2000;238(9):765–773.
11. Narvaez J, Cherwek DH, Stulting RD, et al. Comparing immersion ultrasound with partial coherence interferometry for intraocular lens power calculation. *Ophthalmic Surg Lasers Imaging.* 2008;39(1):30–34.
12. Hasemeyer S, Hugger P, Jonas JB. Preoperative biometry of cataractous eyes using partial coherence laser interferometry. *Graefes Arch Clin Exp Ophthalmol.* 2003;241(3):251–252. No abstract available.
13. Rajan MS, Keilhorn I, Bell JA. Partial coherence laser interferometry vs conventional ultrasound biometry in intraocular lens power calculations. *Eye.* 2002;16(5):552–556.
14. Tehrani M, Krummenauer F, Kumar R, Dick HB. Comparison of biometric measurements using partial coherence interferometry and applanation ultrasound. *J Cataract Refract Surg.* 2003;29(4):747–752.
15. Verhulst E, Vrijghem JC. Accuracy of intraocular lens power calculations using the Zeiss IOL master. A prospective study. *Bull Soc Belge Ophthalmol.* 2001;(281):61–65.
16. Centers for Disease Control. Recommendations for preventing possible transmission of human T-lymphotrophic virus type III/lymphadenopathy-associated virus from tears. *Morb Mortal Wkly Rep.* 1985;34(34):533–534. No abstract available.
17. Velázquez-Estades LJ, Wanger A, Kellaway J, Hardten DR, Prager TC. Microbial contamination of immersion biometry ultrasound equipment. *Ophthalmology.* 2005;112(5):e13-e18.

Suggested Readings

Kahn HA, Leibowitz HM, Ganley JP, et al. The Framingham Eye Study. I. Outline and major prevalence findings. *Am J Epidemiol.* 1977;106(1):17–32.

Four costliest outpatient procedures. *Hospital Health News.* 1998;72:32–33.

Steinberg EP, Javitt JC, Sharkey PD, et al. The content and cost of cataract surgery. *Arch Ophthalmol.* 1993;111(8):1041–1049.

Holladay JT, Prager TC, Ruiz RS, Lewis JW, Rosenthal H. Improving the predictability of intraocular lens power calculations. *Arch Ophthalmol.* 1986;104(4):539–541.

Javitt JC, Brenner MH, Curbow B, Legro MW, Street DA. Outcomes of cataract surgery. Improvement in visual acuity and subjective visual function after surgery in the first, second, and both eyes. *Arch Ophthalmol.* 1993;111(5):686–691.

Mangione CM, Phillips RS, Lawrence MG, et al. Improved visual function and attenuation of declines in health-related quality of life after cataract extraction. *Arch Ophthalmol.* 1994;112(11):1419–1425.

Prager TC, Chuang AZ, Slater CH, Glasser JH, Ruiz RS. The Houston Vision Assessment Test (HVAT): An assessment of validity. The Cataract Outcome Study Group. *Ophthalmic Epidemiol.* 2000;7(2):87–102.

Axial Length: Laser Interferometry
Basics of the IOLMaster

Wolfgang Haigis, MS, PhD

Zeiss IOLMaster Optical Biometry

The IOLMaster by Carl Zeiss Meditec (Fig. 9-1), introduced in 1999, is an all-in-one device allowing all measurements necessary for the calculation of IOL powers to be performed with one instrument.[1] It includes an automatic keratometer to determine central corneal curvatures as well as a slit-image-based setup to measure anterior chamber depth. The IOLMaster's most important feature, however, is the use of partial coherence interferometry (PCI)—also named laser Doppler interferometry (LDI), optical coherence biometry (OCB), or laser interference biometry (LIB)—to measure AL. PCI biometry was developed by the Austrian physicist A. F. Fercher,[2] who performed the first AL measurement in vivo in 1986.

All IOLMaster measurements are noncontact procedures, easily performed and well accepted by patients. The instrument's operating software includes databases for IOL and surgeon data and offers IOL power calculations with all popular formulas.

The reproducibility for AL measurements with the IOLMaster is 22 µm. The instrument is highly observer-independent, as the inter- and intra-observer variabilities show: AL: 10-12 µm, ACD: 31-38 µm, and corneal radii: 11-14 µm (Fig. 9-2).

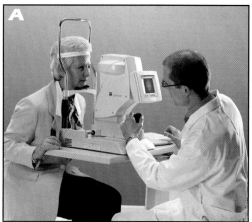
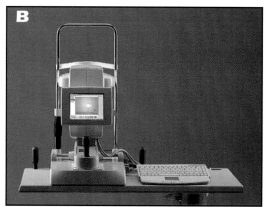
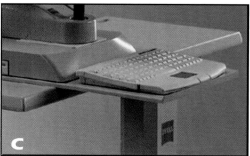

Figure 9-1. Zeiss IOLMaster. (A) side view, (B) examiner view, (C) computer keyboard. (Reprinted with permission from Carl Zeiss.)

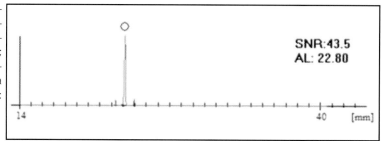

Figure 9-2. AL measurement with IOLMaster (software version 5.x, composite signal): AL = 22.80 mm; SNR = 43.5. The signal-to-noise ratio (SNR) is a measure of signal quality: the higher the better.

Measuring AL with Partial Coherence Interferometry

The AL measuring principle of the IOLMaster is based on dual-beam partial coherence interferometry. The term "coherence" describes the physical property of waves having a temporally constant or regularly varying phase difference at every point in space. Coherence is a necessary requirement for interference. Partially coherent light rays can interfere with each other if they meet within their coherence lengths.

In the IOLMaster, a laser diode emitting partially coherent light (coherence length ≈160 µm) in the near infrared (at a wavelength of 780 nm) is part of a Michelson interferometer setup, which (by means of a moving mirror) produces two partial beams of different optical path lengths. The patient fixates onto the light of the laser diode, thus offering his visual axis to the measuring laser. Both partial beams are reflected at the cornea and at the retina. An interference signal is obtained, eg, when the optical path length of the displacement of the moving mirror in the Michelson interferometer is equal to the optical path length between cornea and retina (ie, the AL of the eye). The optical path length equals the geometrical length times the (group) refractive index of the medium through which the light travels. The position of the interferometer's moving mirror can be measured very precisely determining the accuracy of the AL measurement with the IOLMaster.

Correlation Between Optical and Ultrasound Biometry

It was already mentioned that the IOLMaster measures along the visual axis of the eye. To be more precise, it measures the optical path length from the anterior corneal vertex to the retinal pigment epithelium (RPE) along the visual axis. Ultrasound, on the other hand, measures from the anterior corneal vertex to the internal limiting membrane (ILM) along the eye's optical axis, because it is this axis which allows all interfaces to be met at right angles thus producing a good A-scan echogram. Since clinical experience built-up for decades is based on ultrasound biometry, it was necessary to translate optical AL measurements into equivalent ultrasound readings. Therefore, the IOLMaster was internally calibrated against high precision immersion ultrasound.[3] Consequently, the IOLMaster displays AL results as if it were a highly precise A-scan instrument performing segmental immersion measurements with laser precision.

As a consequence of this equivalence between IOLMaster and immersion AL, it was necessary to adapt the existing IOL constants, which are given by lens manufacturers and usually meant to be used for contact ultrasound. For most of today's IOLs, optimized constants can be downloaded directly into the IOLMaster from the website of the User Group for Laser Interference Biometry (ULIB).[4]

References

1. Haigis W. Optical coherence biometry. In: Kohnen T, ed. *Modern Cataract Surgery. Dev Ophthalmol.* Basel: Karger; 2002;34:119–130.
2. Fercher AF, Roth E. Ophthalmic laser interferometer. *Proc SPIE.* 1986;658:48–51.
3. Haigis W, Lege B, Miller N, Schneider B. Comparison of immersion ultrasound biometry and partial coherence interferometry for intraocular lens calculation according to Haigis. *Graefe's Arch Clin Exp Ophthalmol.* 2000;238(9):765–773.
4. ULIB, User Group for Laser Interference Biometry. http://www.augenklinik.uni-wuerzburg.de/ulib/index.htm (accessed April 1, 2010).

10

IOLMaster Examination

Wolfgang Haigis, MS, PhD

General Information

1. Take optical measurements first, prior to any ocular examination requiring eye contact or application of drops.
2. Check the calibration of the IOLMaster daily before starting measurements on patients.
3. Adjust the instrument table, the headrest, and the IOLMaster so that the patient sits in a relaxed yet stable position.
4. Explain the measurement procedure and point out the necessity for the patient's head to remain in a fixed position with no unnecessary eye movements.
5. Let the patient blink before starting measurements and ask them to focus steadily on the fixation light. Make sure that the patient sees the fixation light.

While focusing is not mandatory with the IOLMaster for AL measurements, it is required for keratometry (K) and ACD measurements. From software Version 5 onwards, adjustment aids are implemented that signal the quality of focusing in a traffic-light manner and automatically start the measurement sequence once focus is optimum.

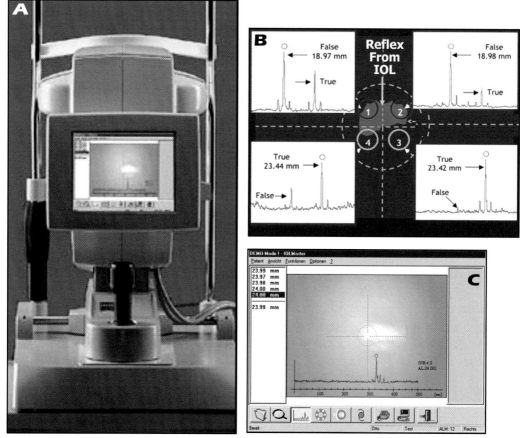

Figure 10-1. Zeiss IOLMaster in AL mode. (A) Instrument from examiner's view, (B) AL screen—note AL readings collected on left side, (C) example of false readings.

Axial Length Measurement

1. Start measurement sequence with a focused and centered beam (Fig. 10-1A and B). If signal quality is bad, try decentering and defocusing.
2. Generally perform at least 5 to 10 measurements per eye. This way, rare artifacts can easily be identified.
3. If the IOLMaster does not display a mean value, which happens when individual measurements differ by more than 100 μm (software Versions 4 and older), check each measurement trace for consistency.
4. Identify the likely origin of inconsistent readings (Fig. 10-1C) (for details refer to the IOLMaster manual). The most frequent signal stems from the RPE. The reason might be a membrane, retinal detachment, or multiple signals from the inner limiting membrane (ILM), RPE, choroid, or artifacts (side lobes to main signal) from the instrument's laser diode.
5. If necessary, zoom in and manually shift (left or right mouse-click) the measurement cursor to the correct signal.

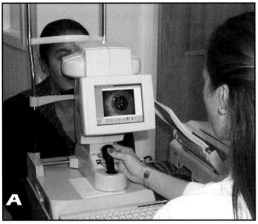 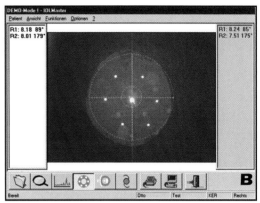

Figure 10-2. Zeiss IOLMaster in K reading mode. (A) Examiner adjusting the reflected mires, (B) K reading screen showing green cross-hairs and reflected corneal mires.

Keratometry

1. Adjust the IOLMaster so that the 6 peripheral light spots, not the one in the center, are in focus (Fig. 10-2A).
2. The eyelid or eyelashes must not obscure these light spots seen in the screen (Fig. 10-2B). If necessary, ask the patient to open their eye widely or assist them in manually keeping it open.
3. Have the patient blink one or more times before triggering the measurement.
4. Perform 3 consistent measurements.

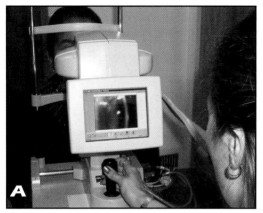 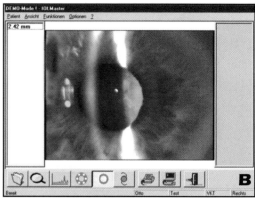

Figure 10-3. Zeiss IOLMaster in ACD mode. (A) Examiner adjusting the AC box and lens surface, (B) ACD screen showing anterior slit beam.

Measurement of Anterior Chamber Depth

1. The image analysis software handling this measurement is designed only for phakic eyes. *Applying it to pseudophakic eyes would produce erroneous results.*
2. Focus on the reflection of the fixation light (the smallest dot) (Fig. 10-3A), and make sure it is within the green square. A slit lamp-type image should be visible (Fig. 10-3B), clearly showing the anterior surface of the crystalline lens inside the pupil. The slit image of the cornea should be well outlined and clear of external reflexes.
3. After successful adjustment, ask the patient to blink before triggering the measurement.

As in ultrasound biometry, the anterior chamber depth displayed is defined from the anterior corneal vertex to the anterior lenticular vertex (ie, including corneal thickness).

If individual measurements differ by more than 100 µm, no mean value will be displayed and the measurement has to be repeated.

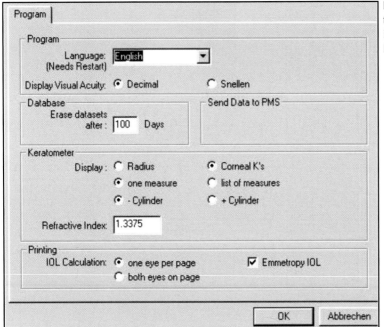

Figure 11-2. Setup screen to adjust keratometer index in software menu of the IOLMaster.

Reference

1. Lege BAM, Haigis W. Laser interference biometry versus ultrasound biometry in certain clinical conditions. *Graefes Arch Clin Exp Ophthalmol.* 2004;242(1):8–12.

Axial Length: Laser Interferometry
The LENSTAR LS900 Instrument

Kenneth J. Hoffer, MD; H. John Shammas, MD; and Jaime Aramberri, MD

Since the introduction of the IOLMaster in 1999, there has been no alternative for optical biometry until 2009, when Haag-Streit (Koniz, Switzerland) introduced their own optical biometer called the LENSTAR LS900. The questions that need to be asked are whether this new instrument:
1. Can match the accuracy and reproducibility of the IOLMaster?
2. Is as easy to use as the IOLMaster?
3. Is as dependent upon dense media and patient fixation ability?
4. Is also superior in staphylomatous eyes and those filled with silicone oil?

Though there have been a few previous studies in the EU comparing the biometry measurements, the authors had access to the first LENSTAR in the US in 2009, and published the report on the IOL power prediction accuracy comparing the two instruments.[1] The LENSTAR provides the surgeon with all axial parameters of the eye (AL, corneal thickness [CT], ACD, and crystalline lens thickness [LT]) on the visual axis, measured with laser interferometry. At the same time, it measures the corneal curvature and axis, white to white distance (corneal diameter [CD]), and pupil diameter and eccentricity of the pupil. All measurements are taken in a single alignment procedure, thus improving the accuracy of the individual measurements since it is all done in the same sweep.

The measurement data is used to predict an IOL using either the internal calculator featuring standard modern formulas (Haigis, Hoffer Q, Holladay 1, and SRK/T) or it can be sent directly to an external IOL calculator featuring 4th generation IOL formulas or

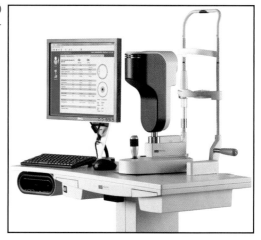

Figure 12-1. Measurement setup. LENSTAR LS900 instrument table with head rest and personal computer.

ray tracing based IOL prediction software. An integrated export interface and a script language tool allow the LENSTAR and its software package (EyeSuite) to be easily connected to any electronic medical record system. DICOM connectivity is planned to be implemented as soon as the respective standards are finalized.

Technical Features

The instrument setup consists of the LENSTAR measurement unit and a standard Windows software computer (Fig. 12-1). This modular concept allows the use of the LENSTAR integrated in varying sizes of its own instrument table or as a stand-alone unit on an instrument table, whereby it fits on a standard Haag-Streit slit lamp table after the slit lamp has been lifted off (Fig. 12-2). One author (Hoffer) has recommended that they simply supply the software on an inexpensive $300 netbook-type computer (Fig. 12-2E), thus eliminating the space needed for the computer, monitor, keyboard, and mouse. Furthermore, it's possible to run additional software like EMR or IOL calculation tools on the same unit and it can also be part of a network.

The measurement unit (Fig. 12-2D) consists of an optical measurement head and a mechanical cross slide system to align the device. Within the optical measurement head there are components to measure the AL, CT, ACD, and LT based on laser interferometry as well as the corneal curvature, CD, and pupillometry based on digital image analysis.

Optical Low Coherence Reflectometry

The axial measurements are very reproducible because of the optical low coherence reflectometry (OLCR) technology used in the LENSTAR. OLCR is a laser interferometric measuring method. It is based on a Michelson interferometer (Fig. 12-3) and is powered by a superluminescence diode (SLD), a broad band light source, with a spectral width of approximately 25 nm, which in turn is centered at 820 nm.[2] The light is distributed to the interferometer with one measurement and two reference arms. Each reference arm consists of an optical single mode fiber, a focusing lens, a mirror, and a rotation cube acting

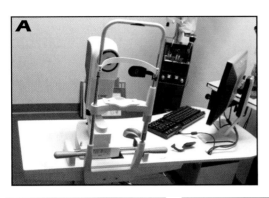

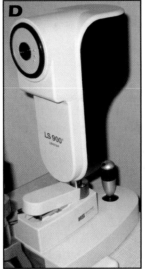

Figure 12-2. Different office setups for the LENSTAR. (A) Full LENSTAR table setup. (B) Slit-lamp setup with computer components on a side table. (C) LENSTAR unit replacing the slit lamp on its table. (D) Close-up photo of the LENSTAR measuring unit. (E) LENSTAR mounted on a Haag-Streit stand attached to a simple laptop computer with the software installed.

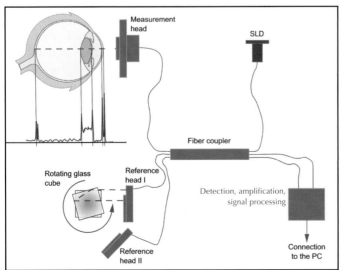

Figure 12-3. Schematic of the optical low coherence reflectometer measurement setup.

Figure 12-4. A-scan of a standard cataract patient measured 5 times. Every A-scan is derived from 16 single scans, using advanced digital signal processing to improve the signal to noise ratio to better penetrate the cataract.

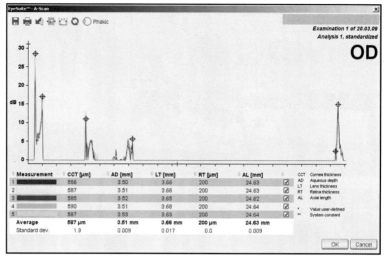

as an optical path length modulator. The reference beam is retro-reflected from the mirror back into the fiber. The measurement arm consists of a sample arm fiber, polarization controllers, and several lenses.

The position of the patient's cornea is monitored by an infrared camera, and the user adjusts the centering on a live video image. Because of the refractive index changes at every interface (eg, air to cornea, cornea to aqueous, etc) the measurement beam is reflected back into the sample arm fiber. An interferogram occurs only when the difference between the optical path lengths of the reference and the sample arm is less than the coherence length of the SLD. Multiple scans are collected, digitally processed, and displayed as a single measurement on the computer screen connected to the LENSTAR measurement unit.

Measurements

Axial measurement of all parts of the human eye is an automated objective measurement. All parameters are measured along the visual axis as the patient fixates on the measurement beam and the unit detects loss of fixation and stops the measurement procedure until the patient re-fixates on it again. Depending on the age, alertness, and cooperativeness of the patient, this process can vary from very easy to tedious or impossible, similar to but more so than the IOLMaster. But in contrast with the latter, once the reading has been taken successfully all the parameters have been measured at one time and there is no need to now measure the K reading, ACD, and CD in separate steps.

A unique feature of the LENSTAR is being able to measure all axial parameters in a single scan but also to display the data in an "A-scan" similar to the displays of A-scan ultrasound biometers. The abscissa refers to the AL and the ordinate to the signal to noise ratio (SNR). All measurements are shown in a single graph (Fig. 12-4), allowing the user to easily assess the quality of the repeated measurements as well as the automatic detection of the individual structures within the eye. If the user does not agree with any of the automatic measurements, they can adjust the gauges in the "A-scan" to measure the structures to their best judgment. User defined measurements are marked on the screen with an asterisk.

Parallel to the axial measurements, the LENSTAR provides the user with measurements of the corneal curvature, CD, and pupil diameter, as well as the eccentricity of the visual axis with respect to the pupil and the CD center. The central corneal curvature is assessed based on two rings of a total of 32 infrared LED markers projected onto the cornea with a 1.65 and 2.3 mm optical zone. Image analysis algorithms are used to derive the main meridians from the reflections, captured by a high-sensitivity CMOS camera. The corneal curvature is displayed either in mm radius of curvature (r) or D of corneal power, for which the user can select his preferred index of refraction to convert r values into D.

The same image is used to derive the pupil diameter of the eye at the time of measurement. Adjusting the ambient light, this measurement feature may be used to estimate the effectiveness of a multifocal IOL on a specific patient.

The CD can either be analyzed on the infrared images taken for the determination of the corneal curvature or from an optionally taken red-free image. This red-free image provides improved contrast as compared to the infrared image, thus improving the repeatability of the measurement. The high contrast red-free image of the eye may also serve as a base for the preoperative planning of a toric IOL, determining landmarks to later orient the IOL in the OR.

Measurement Procedure

The concept of having the measurement unit and the computer separate allowed the creation of a graphical user interface. A measurement wizard guides the user through the measurement procedure and supports their interaction with the patient by providing clear text messages of how to improve the measurement results, eg, if the eye lashes of a patient block marker points of the keratometry measurement.

The starting point of every measurement is the selection of the measurement mode. The measurement modes are phakic, aphakic, pseudophakic, and silicone-filled eye status. In the pseudophakic mode it is possible to select the appropriate IOL material. The option silicone-filled eye can be combined with any of the other measurement modes. If the user accidentally forgets to select the correct measurement mode, it is possible to switch again later in the analysis of the "A-scan."

The next step is the alignment of the device on the patient's eye. First the user coarsely aligns the measurement beam in the center of a cross-hair displayed on an overview live image of the eye on the computer screen, and then focuses on the reflections of the keratometry measurement points. When centered properly, the thumb button is then pressed on the joystick and a zoomed live image is seen and then the user performs the fine centering and focusing—which is supported by online feedback of the device. As soon as the measurement position is reached, a green circle is displayed (Fig. 12-5). The size of the circle represents where in the measurement range the LENSTAR is positioned. Originally when the circle was large it was red and when medium it was yellow. Because this gave the impression that the examiner had to keep focusing until obtaining the green (smallest) circle, one author (Hoffer) recommended that they eliminate the red and yellow colors, which they have done. The smaller the circle, the better the device is centered in the measurement range. The measurement quality is independent of the position of the device in the measurement range of 3 mm in the direction away from and closer to the patient.

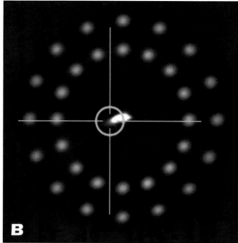

Figure 12-5. (A) Measurement wizard and result overview at the time measurements are taken. The arrow in the measurement screen guides the user to move the LENSTAR unit away from the patient to center the device in the measurement range. (B) Close-up of the screen image of the aiming target.

Figure 12-6. Results screen with cross-sectional and frontal schematic of the human eye to educate the patient on the measurements just taken.

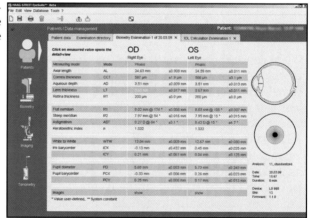

As soon as the measurement range is reached, one may press the joystick button again to start the measurement, an automatic process. During the measurement, the LENSTAR detects blinking and loss of fixation. In such cases, the measurement procedure is interrupted and continued after the patient regains fixation. During the measurement, a grey circle starts at the 12:00 position and completes a 360° circle in a clockwise direction. This represents the advancement of the "A-scan" and biometry data acquisition. The cycle completes when the 16 scans and 4 K readings have been acquired. Collecting so much more information obviously makes this step slightly longer than with the IOLMaster. This was truer with the original software (Version 1.04) we used, but is much faster with the latest software (Version 1.10). With the latter, a novice user took 3 minutes to obtain a result while an experienced user took only 90 seconds.

The result of every scan is immediately displayed in a summary and also as a small "A-scan." It is strongly recommended to take 5 consecutive measurements to get a valid statistical feedback. After all measurements on both eyes have been taken, the results may be reviewed on the summary screen (Fig. 12-6). To easily educate the patient on the measurements just taken, this screen shows a schematic cross section and frontal view of an eye, showing the individual measurements as the operator moves the mouse over the respective values. Clicking on the values shows the details similar to an A-scan and enables an easy check of the measurement quality and, if required, manual adjustment to the user's best judgment. At this stage the measurement procedure is completed. The number of measurements taken on a single patient per day is not limited. The light intensity of the SLD emitted to the patient would allow more than eight hours of uninterrupted measurement without causing any harm to the eye.

IOL Calculation

Besides the measurement procedure, the IOL calculation screen is called the EyeSuite Biometry (Haag-Streit USA, Mason, Ohio). It provides an integrated IOL calculation module. The user may create their own individual set of IOLs and prediction formulas. Per sheet, a combination of 5 IOL/formula combinations is possible. The integrated IOL calculation formulas are Haigis, Hoffer Q, Holladay 1, SRK/T, and (unfortunately) SRK II. The IOL calculation module may be used in a network environment independently of the measurement unit. Recalculation while in the operating room is also possible.

Apart from the integrated IOL calculation, EyeSuite's export interface allows connection to external IOL prediction software such as Dr. Hoffer's Hoffer Programs® (SureID, Mullica Hill, NJ), Dr. Holladay's Holladay IOL Consultant® (Holladay Consulting Inc, Bellaire, TX), Dr. Olsen's PhacoOptics® (IOL Innovations Aps, Aarhus N, Denmark), and Dr. Preussner's Okulix® (der Leu, Hillerse, Germany). Other software might easily be adapted to EyeSuite creating interfaces based on the Haag-Streit script language.

Accuracy Studies

A clinical study performed at the University Clinic Insel, Berne, Switzerland by Rohrer et al[3] on 144 eyes found very good repeatability of the AL measurement as displayed in Table 12-1, where data for patients with a complete set of 5 repeat measurements for both eyes were included to derive mean and standard deviation (SD) and the coefficient of variation (CV). For each measurement parameter, the SDs of each set of 5 replicate measurements were found to be nearly constant over the range of mean results. This finding confirms that the indication of one single SD value for each measurement parameter is a valid approach. The study not only included standard cataract patients but also patients who had previously undergone lens extraction and/or vitrectomy. Besides demonstrating excellent repeatability, the study also showed very good agreement of the LENSTAR measurements with those of the IOLMaster.

Corneal curvature was also analyzed in the clinical study by Rohrer et al. Again, good repeatability and agreement with the IOLMaster was reported. These findings are in agreement with studies by Rabsilber et al,[4] Holzer et al,[5] Buckherst et al,[6] and Cruysberg et al.[7] Even though some of these studies showed statistically significant differences in some of the parameters measured between the LENSTAR and the IOLMaster, these differences did not have any significant impact on the IOL calculation using IOLMaster optimized lens constants from the User Group for Laser Interferometric Biometry[8] (ULIB) on both devices.

Table 12-1.

DATA FOR SUBJECTS WITH A COMPLETE SET OF 5 REPEAT MEASUREMENTS FOR BOTH EYES TOTAL MEAN, THE STD DEV REPEAT AND THE COEFFICIENT OF VARIATION (CV)

Measurement	$Mean_{grand}$	$Std\ Dev_{repeat}$	CV
AL (mm)	23.973	±0.035	0.00145
Corneal thickness (µm)	557.100	±2.300	0.00407
ACD (mm)	3.190	±0.040	0.01220
Lens thickness (mm)	4.560	±0.080	0.01784
Corneal curvature (mm)	7.670	±0.030	0.00396
Axis of the steep meridian (°)	162.000	±11.000	0.14191
White to white distance (mm)	12.270	±0.040	0.00337

The study by Hoffer, Shammas & Savini[1] showed excellent comparison between the LENSTAR and IOLMaster (Table 12-2). Fifty eyes with clear lenses and 50 cataractous preoperative eyes were measured performing the exam on only one eye of each patient. Lens thickness could not be compared because it is not measured by the IOLMaster so we compared it to our previous ultrasound studies: 600 eyes[9] and 1000 eyes[10] showing a 0.11 mm thinner reading with ultrasound. The IOL power prediction using the Haigis formula showed practically no clinical difference between the two instruments.

Comparing the Technical Time Involved Using the IOLMaster vs the LENSTAR

IOL power calculation demands the input of different variables depending on which formula is used. Third generation formulas only use AL and K,[11-14] while 4th generation formulas ask for more: ACD, LT, CD, etc.[15,16] Measurement of these parameters has been simplified by the marketing of devices that can measure all or most of them by optical technologies.[17]

Beyond precision and accuracy, the time it takes to perform the measurement is a relevant issue that matters in such frequently performed procedures. Faster devices are positively valued by users that tend to consider faster machines more user-friendly. Speed is dependant upon software and hardware design and therefore significant differences can be found among the presently marketed optical biometers such as the IOLMaster (Zeiss), LENSTAR LS900 (Haag-Streit), Biograph (Wavelight, Erlangen, Germany), OA 1000 (Tomey, Nagoya, Japan), etc.

Aramberri performed a comparative study of the work time of measurement of the IOLMaster (Version 5.4.3) and the LENSTAR LS900 (Version 1.3.0).

The IOLMaster measures AL by Partial Coherence Interferometry (PCI), the ACD by slit image analysis, the K by automated keratometry, and the CD by image analysis. Each

Table 12-2.

BIOMETRY FOR 50 EYES WITH CATARACT AND 50 EYES WITH A CLEAR LENS COMPARING LENSTAR WITH IOLMASTER

Measurement	LENSTAR	IOLMaster	Range of Diff
Biometry of 50 Eyes with Clear Lenses			
AL (mm)	23.72 ±1.21	23.70 ±1.20	-0.05 to +0.11
ACD (mm)	3.10 ±0.41	2.95 ±0.39	-0.24 to +0.40
Corneal power K (D)	43.41 ±2.13	43.53 ±2.13	-0.80 to +0.33
Biometry of 50 Eyes with Cataracts (Pre-op)			
AL (mm)	23.71 ±1.04	23.68 ±1.04	-0.05 to +0.14
Corneal thickness (μm)	557.100 ±2.3	NA	NA
ACD (mm)	3.11 ±0.47	2.98 ±0.49	-0.22 to +0.55
Lens thickness (mm)	4.72 ±0.47 3.76 to 6.50	NA	*US 4.63 ±0.68 **US 4.63 ±0.46
Corneal power K (D)	43.58 ±1.87	43.69 ±1.92	-0.61 to +0.34
IOL Power Calculation Prediction Error (Haigis Formula) 50 Eyes			
Mean Error (ME)	-0.002 ±0.56	+0.003 ±0.55	
Mean Absolute Error (MAE)	0.455 ±0.32	0.461 ±0.31	
Range of Error	-1.06 to +1.38	-1.17 to +1.13	
Error ±0.50 D	58%	56%	
Error ±1.00 D	94%	94%	
Error ±1.50 D	100%	100%	

*Hoffer US study of 600 eyes[9]
**Shammas US study of 1000 lenses[10]

parameter is measured individually and the user has to go consecutively through different screens using a laptop-style touchpad, focusing the target with a joystick and clicking to obtain each measure. In AL mode each click obtains 1 measurement. This software version demands at least 5 measurements to calculate an average result with a composite SNR value. After this the user enters into the K mode, focuses, and with one click 3 measurements are obtained. Then the ACD mode is selected, and the image is focused and another single click takes 5 measurements. Then the CD mode can be entered, and, as in AL mode, each click obtains 1 measurement that has to be confirmed once the user checks it in the screen. It is very easy to use and the process of focusing in each mode is fast, taking just a few seconds even for inexperienced users.

The LENSTAR LS900 measures all optical interfaces from anterior cornea to retinal pigment epithelium by means of Optical Low Coherence Reflectometry (OLCR). The eye is segmented in the same way ultrasound A-scan does, so the user finds a familiar graphic display with peaks at each optical medium boundary. The measuring process is very simple as all measurements are obtained from just one click. The user clicks twice to start focusing on a zoomed image of the pupil center. Then another click starts the measuring phase that takes some seconds until the results are displayed on the screen. To get the CD, another click is needed to process the data, which takes another couple of seconds. The LENSTAR LS900 is considered as user-friendly as the IOLMaster by users and allows correct measuring the first time a new examiner tries it.

Clinical Comparison

Both devices were compared in a clinical setting with the aim of determining which unit is the fastest in the best case scenario. Sixteen eyes of 16 hospital staff and non-cataract patients were measured by two expert technicians used to working with all types of ophthalmic diagnostic equipment. There is a source of bias, since they have had 6 years of experience with the IOLMaster and only 1 month with the LENSTAR. The patients were free of any ocular disease and were able to maintain fixation for the examination.

Four situations were defined for comparison:
1. 1 measurement of AL+K+ACD
2. 1 measurement of AL+K+ACD+CD
3. 3 measurements of AL+K+ACD
4. 3 measurements of AL+K+ACD+CD

CD was only included in 2 cases, as this is a parameter only used by the Holladay 2 formula and many surgeons don't use it routinely. One measurement was selected as reference and 3 measurements as a common strategy used by many surgeons to average the final parameters, although we are aware that 5 measurements is recommended to ensure precision.

Results

The results of the time measurements can be seen in Table 12-3 and Fig. 12-7. The IOLMaster measures consecutively AL, K, and ACD in 15.56 seconds (±1.36). If CD is added another 3.5 seconds are needed, on average. Scaling the number of measurements to 3 only means a little increase in time as only AL and CD measurements should be repeated with this software version. Three AL, K, and ACD measurements are obtained in 22.06 seconds (±1.69) while 32.38 seconds (±1.93) are necessary to add CD to the results set. The LENSTAR can measure faster if only 1 measurement is done. It takes 12.19 seconds (±0.98) for AL, K, and ACD and 14.19 seconds (±0.98) if CD is also measured. However, increasing the number of measurements means repeating the entire process, so this mean value is multiplied by the number of measurements to be performed. This penalizes this device, making it much slower than the IOLMaster in a multiple measurement setting. The higher the number of measurements, the greater the difference between them. All differences were significant ($p<0.001$) with the Wilcoxon test (Table 12-4).

Table 12-3.

DESCRIPTIVE STATISTICS OF TIME (IN SECONDS) NEEDED TO MEASURE AL, K, ACD AND CD

Instrument	Measures	N	Mean	SD	Min.	Max.
IOLMaster	3 AL+K+ACD	16	22.06	±1.69	20.00	27.00
	3 AL+K+ACD+CD	16	32.38	±1.93	29.00	38.00
	1 AL+K+ACD	16	15.56	±1.36	14.00	20.00
	1 AL+K+ACD+CD	16	19.06	±1.61	17.00	24.00
LENSTAR	3 AL+K+ACD	16	36.50	±2.50	32.00	42.00
	3 AL+K+ACD+CD	16	42.50	±2.50	38.00	48.00
	1 AL+K+ACD	16	12.19	±0.98	11.00	15.00
	1 AL+K+ACD+CD	16	14.19	±0.98	13.00	17.00

N = number of eyes, SD = standard deviation

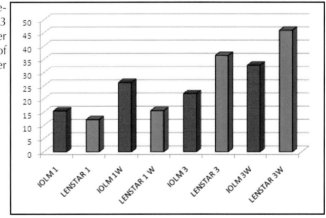

Figure 12-7. Mean time of measurement. IOLM = IOLMaster; 1 or 3 measurements; W = CD parameter measured. Increasing the number of measurements makes LENSTAR slower than IOLMaster.

Table 12-4.

WILCOXON SIGNED RANK TEST WAS USED FOR PAIRED COMPARISON AMONG VARIABLES

Test Statistics	3 AL+K+ACD	3 AL+K+ACD+CD	1 AL+K+ACD	1 AL+K+ACD+CD
Z	-3.526	-3.523	-3.493	-3.457
Asymp. Sig. (2-tailed)	0.000	0.000	0.000	0.001

It´s interesting to observe that variance is smaller with the LENSTAR when only 1 measurement is done, and this is not affected if CD is added to the data set. However, repeating measurements increases dispersion more than with the IOLMaster. Analyzing (case to case) shows that the main contributor to the IOLMaster variance is ACD measurement, whose focusing process can take more time than the one needed for getting AL, K, or CD.

Conclusion

Both devices are equally fast and user friendly. However, some differences can be shown. The LENSTAR is faster for one measurement, but clearly becomes slower than the IOLMaster as more measurements are obtained. However, the magnitude of 10 to 20 seconds per eye is not clinically relevant, even in a high volume setting. In order to improve these numbers, manufacturers should work toward decreasing the focusing time as well as the number of measurements required when performing multiple measurements.

Summary

Measurement of all optical structures in the human eye:
- AL
- Corneal thickness (CT)
- ACD (aqueous depth (AQD) + CT)
- Lens thickness (LT)
- Corneal curvature (main meridians and axis position)
- Corneal Diameter (CD)
- Pupillometry

Eccentricity of the visual axis with respect to CD and the pupil center:
- Complete "A-scan" of the eye
- All measurements taken simultaneously on the visual axis
- Automatic measurement procedure with fixation and blinking control
- Interactive measurement wizard
- Integrated IOL calculation tool
- Interface to EMR systems and third party IOL calculation software

Networkable with other instrumentation and software (DICOM)

References

1. Hoffer KJ, Shammas, Savini G. Comparison of two optical biometers. *J Cataract Refract Surg.* 2010;36(4):644-648.
2. Ballif J, Gianotti R, Chavanne P, et al. Rapid and scalable scans at 21 m/s in optical low-coherence reflectometry. *Opt Lett.* 1997;22:757-759.
3. Rohrer K, Frueh BE, Wälti R, et al. Comparison and evaluation of ocular biometry using a new noncontact optical low-coherence reflectometer. *Ophthalmol.* 2009;116:2087-2092.

4. Rabsilber TM, Jepsen C, Auffarth GU, Holzer MP. Intraocular lens power calculation: Clinical comparison of 2 optical biometry devices. *J Cataract Refract Surg.* 2010;36:230-234.
5. Holzer MP, Mamusa M, Auffarth GU. Accuracy of a new partial coherence interferometry analyzer for biometric measurements. *Br J Ophthalmol.* 2009;93:807-810.
6. Buckhurst PJ, Wolffsohn JS, Shah S, et al. A new optical low coherence reflectometry device for ocular biometry in cataract patients. *Br J Ophthalmol.* 2009;93:949-953.
7. Cruysberg LP, Doors M, Verbakel F, et al. Evaluation of the LENSTAR all-in-one non-contact biometry meter. *Br J Ophthalmol.* 2010;94:106-110.
8. User Group for Laser Interference Biometry, University of Würzburg. http://www.augenklinik.uni-wuerzburg.de/ulib/ (accessed April 1, 2010).
9. Hoffer KJ. Axial dimension of the human cataractous lens. *Arch Ophthalmol.* 1993;111:914-918, erratum 1626.
10. Shammas HJ. A-scan biometry of 1000 cataractous eyes. In: Ossoinig KC, ed. *Ophthalmic Echography.* Dordrecht, the Netherlands: Junk Publishers; 1987:57-63.
11. Hoffer KJ. The Hoffer Q formula: A comparison of theoretic and regression formulas. *J Cataract Refract Surg.* 1993,19:700-712, errata 1994:20:677 and 2007;33:2-3.
12. Retzlaff JA, Sanders DR, Kraff MC. Development of the SRK/T intraocular lens implant power calculation formula. *J Cataract Refract Surg.* 1990;16:333-340, erratum 1990;16:528.
13. Holladay JT, Praeger TC, Chandler TY et al. A three-part system for refining intraocular lens power calculations. *J Cataract Refract Surg.* 1988;14:17-24.
14. Haigis W. *IOL calculation according to Haigis.* 1997. http://www.augenklinik.uniwuerzburg.de/uslab/ioltxt/haie.htm (accessed on April 1, 2010).
15. Olsen T, Oleson H, Thim K. Prediction of postoperative intraocular lens chamber depth. *J Cataract Refract Surg.* 1990;16:587-590.
16. Holladay JR. Achieving emmetropia in extremely short eyes with two piggyback posterior chamber intraocular lenses. *Ophthalmol.* 1996;103:1118-1123.
17. Findl O, Drexler W, Menapace R, et al. High precision biometry of pseudophakic eyes using partial coherence interferometry. *J Cataract Refract Surg.* 1998;24:1087-1093.

13

Corneal Power: Diopters Versus Radius

Wolfgang Haigis, MS, PhD

Definition of Corneal Power in Gaussian Optics

About two-thirds of the total refractive power of the eye is provided by the cornea. In the most basic approximation (Equation 1), the cornea can be represented by a meniscus lens with refractive index n, confined anteriorly and posteriorly by spherical surfaces with radii Ra (anterior) and Rp (posterior) and separated by d (corneal thickness). With $n1 = 1.000$ being the refractive index of air and $n2 = 1.336$ the refractive index of aqueous, each corneal surface can be characterized by its surface powers Da and Dp:

(1) $\quad Da = \dfrac{n - n1}{Ra} \quad$ and $\quad Dp = \dfrac{n2 - n}{Rp}$

In paraxial (Gaussian) optics (Fig. 13-1), the refractive power of this lens can be expressed either as total (equivalent) power De or as back vertex power Dv given by Equation 2[1]:

(2) $\quad De = Da + Dp - \dfrac{d}{n} \cdot Da \cdot Dp \quad$ and $\quad Dv = \dfrac{De}{1 - \dfrac{d}{n} \cdot Da} = Dp + \dfrac{Da}{1 - \dfrac{d}{n} \cdot Da}$

With the respective data of the Gullstrand eye, the refractive powers of the Gullstrand cornea in Table 13-1 are obtained.

Figure 13-1. Cornea in paraxial approximation.

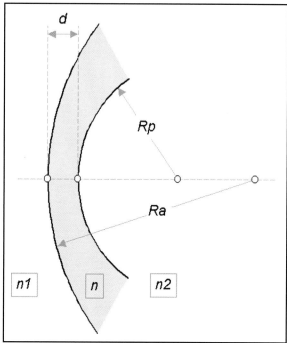

Table 13-1.							
DATA FOR THE CORNEA OF THE GULLSTRAND EYE							
Ra (mm)	Rp (mm)	D (mm)	n	Da (D)	Dp (D)	De (D)	Dv (D)
7.7	6.8	0.5	1.376	48.83	−5.88	43.05	43.83

Measurement of Corneal Power

There is no instrument to directly measure the dioptric power of the cornea. Topographic instruments, keratometers, and ophthalmometers do not measure diopters. Topographic instruments determine radii of curvature from elevation maps, and keratometers and ophthalmometers derive the anterior corneal radius of curvature from the size of the fist Purkinje reflex of a luminant object in the instrument.

The translation from the measured radius Ra in millimeters into a refractive power K (K-reading) in diopters is done by instrument-specific formulas and keratometer indices n'. Among them are the refractive, axial, and instantaneous formulas.[2] The instantaneous is used most and given by Equation 3:

$$(3) \quad K = \frac{n'-1}{Ra}$$

which has the form of a surface power (Equation 1). Differences exist in the numerical value of the keratometer index n'. Javal-Schiötz-type keratometers and all topographers use $n' = 1.3375$, whereas Zeiss instruments in Europe prefer a value of 1.332. In commercial instruments, n' values from 1.3315 to 1.338 can be found. The Javal calibration (1.3375) is close to the refractive index of the tear film, aqueous, and vitreous (1.336), for which Helmholtz had derived a value of 1.337. This value was converted to 1.3375 to give a $K = 45.00$ D for a corneal radius of 7.50 mm, merely for convenience. The same radius produces $K = 44.20$ D with a keratometer index of 1.3315.

According to Gullstrand, the true refractive index of the cornea is 1.376 (see Table 13-1). The K value obtained from Equation 3 with this value is equivalent to the anterior surface power.

Special Meaning of Keratometer Indices 1.3315 and 1.3375

The cornea's total (equivalent) power *(De)* as well as its back vertex power *(Dv)*, according to Equation 2, depend on its front and back surface powers. But keratometry and topography instruments measure only the anterior radius of curvature, *Ra*. Without knowledge of the posterior radius *(Rp)*, neither corneal power *De* nor *Dv* can be calculated. However, if it can be assumed that anterior and posterior radii have a fixed ratio *Rp/Ra*, then it is possible to derive the back surface properties from the anterior surface data. If this ratio is given by the Gullstrand eye (see Table 13-1), ie, $Rp/Ra = 6.8/7.7 = 0.883$, then it can be shown[1] that Equation 3, with an index of 1.3315, gives the total (equivalent) power *De*, with an index of 1.3375, the back vertex power *Dv* of the cornea.

So with radii given in millimeters and refractive powers in diopters, the following formulas hold for Gullstrand-like eyes with a ratio $Rp/Ra = 6.8/7.7$ (Equations 4 and 5):

(4) Back vertex power: $K = \dfrac{337.5}{Ra}$

(5) Total equivalent power: $K = \dfrac{331.5}{Ra}$

An important consequence of Equations 4 and 5 is in effect in eyes after corneal refractive surgery: here, the ratio *Rp/Ra* has been deliberately altered. Therefore, K-readings based on Equations 4 and 5 have no physiological or optical meaning.

The corneal radii of a patient measured with different keratometers or topographers should be the same or differ only slightly, whereas the corneal power may be expected to vary significantly for the different instruments, depending on the respective keratometer index used.

References

1. Haigis W. Corneal power after refractive surgery with myopia: the contact lens method. *J Cataract Refract Surg.* 2003;29(10):1397–1411.
2. Roberts C. The accuracy of "power" maps to display curvature data in corneal topography systems. *Invest Ophthalmol Vis Sci.* 1994;35(9):3525–3532.

Corneal Power: Manual Keratometry and Instrumentation

Kenneth J. Hoffer, MD

Manual keratometry is a method to determine the true central optical power of the cornea by measuring the radius of curvature of the anterior surface. It makes assumptions regarding the posterior surface (which cannot be measured by the instrument) and then converts the radius into diopters (D) by a simple equation:

$$D = 337.5/r$$

where D = diopters, r = radius of curvature.

Arguments rage as to what the index of refraction should truly be, which is not the subject of this chapter.

The original instrument for measuring corneal power manually is called a keratometer or an ophthalmometer. The principle by which the instrument measures the anterior corneal curvature is contingent upon accurately determining the size of a reflected image from the front surface of the cornea or, as it is called, the first Purkinje-Sanson image.

In manual keratometers the examiner must align and focus the illuminated mires (Fig. 14-1), as well as modify their position to get the measurements needed. All keratometers measure the corneal power of the central approximately 3 mm area (from 2.8 to 3.2 mm in unoperated corneas). A problem arises in post-refractive corneal surgery eyes because the central 3 mm area is so much flatter that the image measured is larger than 3 mm, causing an error in obtaining the true "central curvature."

The keratometer projects an image (see Fig. 14-1) onto the cornea and then measures the size of that image reflected from the corneal surface. The device then converts image

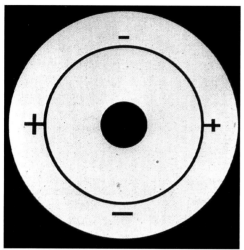

Figure 14-1. Illuminated mires of the keratometer that are projected onto the cornea.

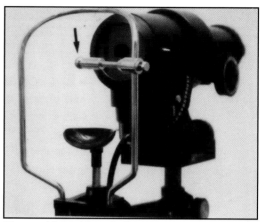

Figure 14-2. Manual keratometer showing the measuring ball in place on the magnetic holder for calibration.

size into corneal radius using simple vergence relationships of convex mirrors. Because of the small but continuous movement of the patient's eyes, the keratometer doubles the reflected image and measures the image against itself, rather than against a fixed scale. There are several commercial models, all sharing some common features:
1. There are two illuminated mires (objects) of known size, located at a set distance from the corneal plane.
2. An image duplicator (doubling device) to make it easier to measure a moving target.
3. An optical system which includes a telescope with a short focal length. This enables the observer to see the resultant image (which is small and virtual).

The various models are divided into those based on the Helmholtz design and those based on the Javal-Schiotz design.
1. In the *Helmholtz* model, the radius of curvature is determined from a known object size (mires) and the size of the object is fixed (separation between the illuminated mires). The size of the image is variable and is obtained by a variable doubling device, a disc with four circular apertures. One of the apertures has a prism that displaces the image in a horizontal plane, while another aperture has a prism that displaces the image in the vertical plane. The variabilility of the system is based upon the possibility of displacing the disc following the axis of the horizontal and vertical prisms until an adequate image is obtained.
2. The *Javal-Schiotz* is based on a fixed image, where the doubling system is not variable, but the size of the object is variable.

The Examination

The patient's chin is placed in the chin rest of the instrument (Fig. 14-2) and the forehead is pressed against the top for fixation. The keratometer housing is then aimed at the eye to be measured and the chin rest and the vertical setting on the housing is then

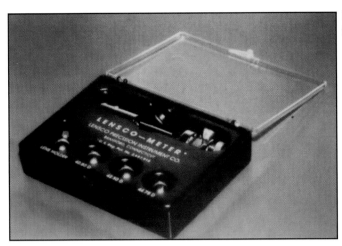

Figure 14-3. Set of calibration balls for the B+L keratometer.

adjusted, after loosening the knobs, so that the reflected image is cast on the center of the cornea. It is often helpful to shine a pen light through the ocular to help grossly align it on the eye. The knobs are then tightened to lock it in place. The horizontal and vertical images are then aligned by rotating the gauges on the left and right of the housing until they are lined up precisely. The readings of corneal power are then taken from the horizontal and vertical gauges. If the values are exactly the same, there is no astigmatism. If there is astigmatism, it is necessary to rotate the whole housing so that the instrument is aligned in the axis of the cylinder and then the gauges are readjusted. The cylinder axis is read from the gauge.

Important Reminders

It is prudent to establish a routine of calibrating manual keratometers using the set of steel calibration balls supplied by the manufacturer (Fig. 14-3). Finding where they are stored may be a problem. When recording corneal power in IOL calculation it is wise to always record the average K, thus ignoring the cylinder which has no effect in these calculations. This eliminates transcription errors. If one can show by past clinical studies that a constant change in the corneal power occurs following your cataract technique, it might be wise to add that effect to all your preoperative K readings. If not it should be ignored. I have not been convinced yet that automated keratometers are as accurate on a consistent basis as manual keratometry.

Lastly, it can be a medicolegal problem if contact lenses have not been left out of the eye continuously for 2 weeks prior to the keratometry exam for IOL power. I obtain patient cooperation with this fact by removing only one contact lens for that period of time.

Note: Specifications on instruments were obtained from the respective manufacturers.

Suggested Reading

Holladay JT. Standardizing constants for ultrasonic biometry, keratometry, and intraocular lens power calculations. *J Cataract Refract Surg.* 1997;23(9):1356–1370.

Automated Keratometry for IOL Power Calculation

Jaime Aramberri, MD

Automated keratometers (Fig. 15-1) measure anterior corneal curvature, providing K values (K) that IOL calculation formulas will use for IOL power calculation. In this process most formulas first use the K, with other predictor variables, to predict the IOL plane position estimation (effective lens position [ELP]). Then the K is used again in the optical calculation vergence formula to solve for IOL power.

The main advantage of automated over manual keratometry is simplicity: joystick focusing and two clicks allow accurate and repeatable measurements in a very short time (Fig. 15-2).

Moreover, handheld autokeratometers (Fig. 15-3) allow the measuring of patients in different postures—which can be useful for examination of children under general anesthesia, disabled or mentally retarded patients, etc.

Other advantages of autokeratometry are displayed in Table 15-1.

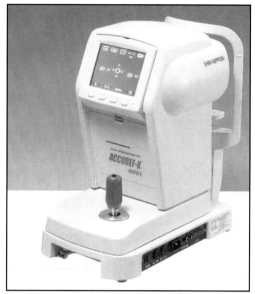

Figure 15-1. A table-mounted autokeratometer.

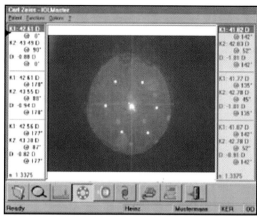

Figure 15-1. A table-mounted autokeratometer.

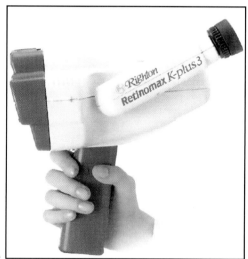

Figure 15-3. A handheld autokeratometer.

Table 15-1.

ADVANTAGES OF AUTOKERATOMETRY

- Examiner-proof method: short learning curve
- Similar accuracy to manual keratometry and videokeratoscopy
- Higher repeatability than manual keratometry or videokeratoscopy
- Fast measuring time: 0.02 to 0.1 seconds
- Eccentricity measurement
- Handheld devices allow measuring non cooperating patients
- Printing of results
- Data interface for a computer network

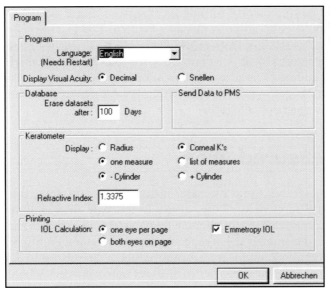

Figure 15-4. Setup screen to adjust the corneal index of refraction in the software menu of the IOLMaster.

Optics

Curvature measurement is based on convex mirror paraxial optics. A known size object is projected onto the cornea from a known distance and the reflection produced in the precorneal tear film is measured. Magnification is related to radius of curvature by a simple paraxial formula:

$$r = 2 d h'/ h$$

where r = radius of curvature, d = object to image distance, h´ = image size, and h = object size.

As measuring speed is high (<0.1 sec), image doubling systems used in manual keratometers are unnecessary.[1] The object mires are different infrared light patterns depending on each machine design: rings, dots, etc.

Measurement

Radius of curvature is usually expressed in mm. Some machines describe the cornea as a spherical toric surface, measuring the steep and flat axis radii which are forced to be 90° apart. Others describe the central cornea as an aspherical toric surface adding an asphericity coefficient to the apical radii.

Conversion from mm to power in D is performed by the paraxial formula:

$$P = n1 - n0 / r$$

where P = power in D, n1 = the index of refraction of cornea, n0 = the index of refraction of air and r = the anterior corneal radius of curvature.

Most keratometers use by default the keratometric standard index of refraction of 1.3375 as n1. This is to compensate for the nonmeasured power of the posterior cornea. Some use other arbitrary indexes, ie, Zeiss uses 1.332. Modern autokeratometers allow changing this parameter by software (Fig. 15-4).

The 1.3375 value has been used in keratometry for historical reasons, but it is known to induce an overestimation of central paraxial power of about 0.75 to 1.00 D. However, the most used IOL calculation formulas (Haigis, Hoffer Q, Holladay, SRK/T) are designed to input that number, correcting it internally. It is essential for the user to know which index is used as it will affect the calculated K reading. A very simple, and strongly advisable, way to avoid conversion errors is getting used to inputting K readings in radius of curvature in mm, eliminating the conversion altogether.

Measured Area

Although most manuals declare a 3 to 3.3 mm diameter measured corneal zone, this figure depends on the curvature and asphericity of the cornea. In this way autokeratometers behave like variable doubling manual keratometers, where object size is fixed. The steeper the corneal anterior surface, the smaller the area reflecting the keratometer mires. An average diameter range extends from 2.8 mm (for a 7 mm radius of curvature) to 3.7 mm (for a 9 mm one).[2] This has clear implications after corneal refractive surgery where steep and prolate or flat and oblate corneas are common.

Performance

The accuracy of autokeratometers has been reported to be at least equal to manual keratometry and videokeratoscopy for K value determination. Repeatability is generally accepted to be better with autokeratometry in normal eyes. Douthwaite et al[3] reported 95% limits of agreement: ±0.033 mm for central radius and ±0.068 for p value measuring conicoidal plastic tests with the Topcon KR-3500 (Topcon Medical Systems, Oakland, NJ). Hammack[4] showed similar results measuring steel calibration spheres with the Bausch + Lomb (B+L) manual keratometer, Nidek ARK-2000 (Nidek Inc, Fremont, CA), Humphrey 410 (Carl Zeiss, Dublin, CA), and Alcon Renaissance (Alcon Laboratories, Fort Worth, TX) autokeratometers. Average and standard deviation of errors were lower than 0.10 D in most cases. Edwards et al[5] proved better repeatability with the Nidek KM-500 and Topcon RK-3000A than with the B+L manual keratometer. Coefficients up to 0.05 mm for the automated devices and up to 0.12 mm for the manual were reported. Giraldez et al[6] found no significant difference in central power among the EyeSys 2000 (EyeSys Vision Inc, Houston, TX) , Javal, and Nidek ARK-700 (maximum mean difference 0.25 D). Ninety-five percent limits of agreement range were clinically significant: ±0.5 D. Sunderraj et al[7] reported a 0.11 D mean difference between the Canon RK-1 (Canon USE, Lake Success, NY) and the Topcon OM-4 manual keratometer. Davies et al[8] compared the Javal keratometer with the NVision-K 5001 (Shin Nippon, Ajinomoto Inc, Tokyo, Japan), finding no bias in central radius and ±0.22 mm 95% limits of agreement for the vertical meridian. There was a significant difference in astigmatism axis; only 45% were within ±10° of the Javal reading. Lam et al[9] reported similar K readings between erect and supine measurements, as long as the handheld keratometer (Nidek KM-500) was correctly aligned with the eye. In this study, difference in central power between topography and autokeratometry was not clinically significant: 0.24 ± 0.78 D. Santodomingo-Rubido et al[10] compared the Zeiss IOLMaster keratometry with the Javal showing a mean difference of -0.03 mm with 0.13 mm 95% confidence limits. They showed with the EyeSys videokeratoscope a mean difference of -0.06 mm with 0.07 mm 95% confidence limits. In this work, repeatability of autokeratometry after an interval of 1 to 10 days was excellent: mean difference was 0.00 mm with 0.04 mm 95% confidence limits.

Autokeratometry has been described to be reliable and not affected by applanation tonometry[11] or gonioscopy. In comparison, videokeratoscopy SimK needed 20 minutes to return to baseline values after gonioscopy both with Goldmann and Sussman lenses.[12]

Sources of Error

The examiner cannot evaluate mire reflection irregularity as in manual keratometry. Moderate irregularity is undetected and can induce measurement error. If irregularity is higher, the device simply doesn't take the reading. Whenever an irregular corneal surface is suspected, videokeratoscopy should be performed to confirm corneal shape.

After corneal refractive surgery K readings will be erroneous, as the corneal anterior/posterior ratio becomes a function of the induced shape change. This must be compensated for if keratometry (manual or automated) is used for K determination for IOL calculation in these eyes.

Conclusion

Automated keratometry is a fast, easy-to-use, accurate, and precise method for corneal central radius measurement. It is advised to use K values in mm instead of D to avoid conversion errors. A list of available autokeratometers is provided in Table 15-2.

References

1. Rabbetts RB, Mallen EAH. Measurement of ocular dimensions. In: Rabbetts RB, ed. *Bennett and Rabbett's Clinical Visual Optics*, 4th ed. Boston: Butterworth-Heinemann; 2007.
2. Stone J. Keratometry and specialist optical instrumentation. In: Ruben M, Guillon M, eds. *Contact Lens Practice*. London: Chapman and Hall Medical; 1994.
3. Douthwaite W, Pardhan S. Accuracy and repeatability of the Topcon K-3500 autokeratometer on calibrated convex surfaces. *Cornea*. 1995;14(3):253–257.
4. Hammack GG. Evaluation of the Alcon Renaissance handheld automated keratometer. *Int Contact Lens Clin*. 1997;24(2):59–65.
5. Edwards MH, Cho P. A new hand-held keratometer: Comparison of the Nidek KM-500 auto keratometer with the B&L keratometer and the Topcon RK-3000A keratometer. *J Br Contact Lens Assoc*. 1996;19(2):45–48.
6. Giraldez MJ, Yebra-Pimentel E, Parafita A, et al. Comparison of keratometric values of healthy eyes measured by Javal keratometer, Nidek autokeratometer and corneal analysis system (EyeSys). *Int Contact Lens Clin*. 2000;27(2):33–40.
7. Sunderraj P. Clinical comparison of automated and manual keratometry in pre-operative ocular biometry. *Eye*. 1992;6(1):60–62.
8. Davies LN, Mallen EAH, Wolffsohn JS, et al. Clinical evaluation of the Shin-Nippon NVision-K 5001/Grand Seiko WR-5100K autorefractor. *Optom Vis Sci*. 2003;80(4):320–324.
9. Lam AKC, Chan R, Chiu R. Effect of posture and artificial tears on corneal power measurements with a handheld automated keratometer. *J Cataract Refract Surg*. 2004;30(3):645–652.
10. Santodomingo-Rubido J, Mallen EAH, Gilmartin B, Wolffsohn JS. A new non-contact optical device for ocular biometry. *Br J Ophthalmol*. 2002;86(4):458–462.
11. Beatty S, Nischal KK, Jones H, Eagling EM. Effect of applanation tonometry on mean corneal curvature. *J Cataract Refract Surg*. 1996;22(7):970–971.
12. George MK, Kuriakose T, DeBroff BM, Emerson JW. The effect of gonioscopy and keratometry and corneal surface topography. *BMC Ophthalmol*. 2006;6:26.

Table 15-2.

COMPILATION OF THE VARIOUS AUTOMATED KERATOMETERS AVAILABLE ON THE MARKET

Manufacturer	Model	Description
Canon, Inc	RK-F1 Auto-Refractor-Keratometer and R-F10 Full Auto-refractor 	Choose between refraction only or with keratometry. One touch does it all: from automatic alignment to precision output. Trackball replaces joystick. Motorized chinrest. Ergonomic controls.
Haag-Streit	OM-900	Uses a classical distance-independent measuring principle, which meets the requirements of Helmholtz. Two different test patterns (Javal or Cross mark) can be used. Can be user-specifically calibrated, thus eliminating subjective systematic measuring errors.
Nikon Instruments, Inc	Speedy-K Auto Refract Keratometer 	Keratometry and refraction readings in 0.35 sec each. Peripheral readings in 0.45 sec each. Medical Retro Illumination Mode. A 100 patient memory. Wider reading range. Easy position adjustment. Video jack. Stopper lever. A 2.5 mm Pupil Diameter Plus Light Intensity Control. Auto shut off.

(continued)

Table 15-2 continued.

COMPILATION OF THE VARIOUS AUTOMATED KERATOMETERS AVAILABLE ON THE MARKET

Manufacturer	Model	Description
Nikon Instruments, Inc	Retinomax K-plus 2 Handheld Auto Refract Keratometer 	Auto-measurement and auto-finish. Selectable reading modes. Fast Ref-Keratometric readings. Shorter reaching distance. Wider reading range. Melody function. Mire ring. Super quick mode. Retro mode key. Peripheral corneal curvature reading. A 2.5 mm pupil size and selectable fixation intensity. Anti-theft design. Flexible configuration.
Marco Ophthalmic, Inc	KM-500 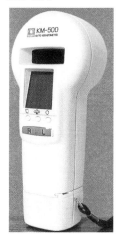	The unique viewing and alignment window lets you observe the targeted eye from any distance. View the eye through the window and move the instrument towards the eye. When the mire ring appears on the cornea, slowly move the unit closer. The KM-500 will fire automatically when properly focused. Readings appear instantly on the LCD display. Both left and right readings display simultaneously. 1.5 lbs. Optional printer. Battery good for one-hour continuous use.
Marco Ophthalmic, Inc	ARK 700A and ARK 760A 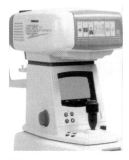	Refractor/keratometer with high-speed mode for readings. Takes 180 separate retinoscopy readings. Auto-fogging to prevent accommodation. IOL button. Automatic alignment, focusing, reading, printing. User-friendly on-screen prompts.

(continued)

Table 15-2 continued.

COMPILATION OF THE VARIOUS AUTOMATED KERATOMETERS AVAILABLE ON THE MARKET

Manufacturer	Model	Description
Marco Ophthalmic, Inc	PALM-AR and PALM – ARK 	Portable. Easy to align. Automatic measurement and fogging. Accurate and dependable readings. Infrared printer communication. Rechargeable battery.
Reichert Ophthalmic Instruments	KR460 Auto Keratometer/Refractor 	Hands-free alignment technology. No joystick or chin-rest adjustments needed. Provides consistent, fast, accurate readings.
Tomey Corporation	RC4000 Auto Refractor-Keratometer 	Multiple functions: Refraction, keratometry, Contact lens base curves, PD. Data output: Built-in printer. Fast and accurate measurements. Power saving feature.

(continued)

Table 15-2 continued.

COMPILATION OF THE VARIOUS AUTOMATED KERATOMETERS AVAILABLE ON THE MARKET

Manufacturer	Model	Description
Topcon Medical Systems, Inc	KR-8000PA SUPRA Auto-Kerato Refractometer 	Combines an auto-refractor, an auto-keratometer and a computerized color corneal mapping system in a single compact unit. Rotary prism technology. Auto-tracking and auto measurement enhance ease of operation. Incorporated placido rings for enhanced testing precision, providing corneal mapping measurements out of 10 mm.
Zeiss Meditec	KR-7000S Auto Kerato-Refractometer 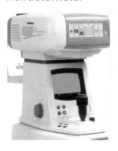	Provides accurate objective refraction, thorough yet simple subjective refinement and keratometry measurements all in one. Online capability to compare refraction to actual spectacle, using computerized lensometer.
Zeiss Meditec	HARK Model 599 	Full featured glare testing. Automatic eye tracking. Immediate patient verification of prescription. Two different configurations fit any space (operator control panel can be positioned at either 90 or 180 degrees from the patient). Complete testing. Automatic keratometric readings. Communicom Interface.

(continued)

Table 15-2 continued.

COMPILATION OF THE VARIOUS AUTOMATED KERATOMETERS AVAILABLE ON THE MARKET

Manufacturer	Model	Description
Zeiss Meditec	Humphrey Acuitus Models 5000 and 5015	Flat screen color display. Thirty measurements per second. Auto XYZ. Expanded refracting capabilities. Auto-acquisition. Full screen eye image. Illuminated icon control buttons. Chinrest and built-in printer. Keratometry only available on model 5015.
Zeiss Meditec 1999	IOL Master	Automatic K readings taken optically. Setup screen allows adjustment of index of refraction used. If not set to 1.3375, the Hoffer Q formula will not report accurate results.
Haag-Sreit Koeniz, Switzerland 2009	LENSTAR LS900 	Automatic K readings taken by laser. Setup screen allows adjustment of index of refraction used. It is not as fast, but in one pass it measures: AL, K, CT, AQD, LT, CD, RT

Some of the instruments listed in this table may no longer be available for new purchase, but may be available on the secondary/used/refurbished market.

Corneal Power: Corneal Topography for IOL Power Calculation

Jaime Aramberri, MD

Corneal topography is the measurement of corneal shape and optical function. It can be useful for IOL power calculation in different ways:

- Determination of corneal central curvature and power for the regular IOL calculation formulas
- More complex IOL calculation using exact ray tracing methods
- Measurement of corneal spherical aberration to calculate aspheric IOLs
- Corneal astigmatism measurement for toric IOL calculation

Other uses in IOL surgery:

- Astigmatism corneal surgery planning and postoperative monitoring
- Corneal diagnosis for postoperative excimer surgery indication

Technologies

Three different technologies can be found in commercial devices (Fig. 16-1).

Figure 16-1. Reflexion topography (Placido videokeratoscopy) measures corneal curvature and calculates elevation. Projection tomography measures elevation and calculates curvature.

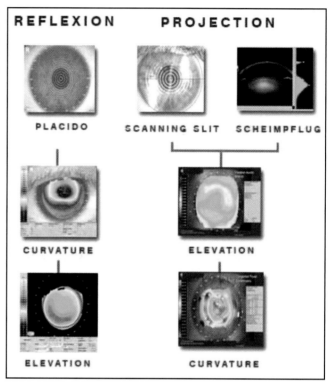

PLACIDO TOPOGRAPHY

Convex mirror optics are used to measure the size of the image reflected by the cornea of a known object at a known distance. The object is usually a set of concentric rings called a Placido disc (Fig. 16-2) so that the flatter the cornea the wider the rings. Curvature reconstruction depends on the algorithm programmed in each topographer. The first models used a keratometric analysis method, calculating shape the way a keratometer does; that is, measuring ring magnification.[1] Modern topographers use different arc step algorithms where a continuous sequence of arcs is traced from the vertex to the periphery of each semi-meridian. This method allows simultaneous calculation of curvature and height.[2] The advantage of the latter is that peripheral slope and curvature, as well as height, are more accurate than with keratometric-like algorithms that incorrectly assume a spherical geometry (spherically biased). Central cornea measurement (SimK) is similar with both reconstruction methods as long as the shape is sphero-cylindrical.[3]

SCANNING SLIT TOMOGRAPHY

The Orbscan IIz (B+L, Rochester, NY) combines Placido disk and scanning slit technology for corneal topography. Forty light slits are projected at 45° angles in 1.5 seconds scanning time. Imaging is performed by a camera aligned with the corneal vertex. Surface elevation is measured, and by differentiation, curvature can be calculated. Anterior corneal topography can be calculated both from scanning slit or from Placido information (Fig. 16-3).

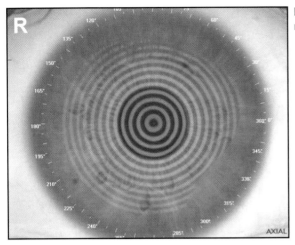

Figure 16-2. Placido disk image showing the reflected rings.

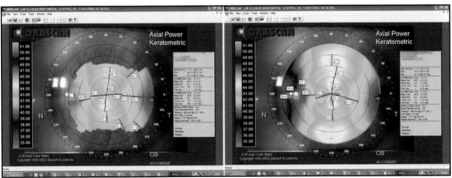

Figure 16-3. Axial maps of the same eye with Orbscan IIz. Left map is generated from Placido data; superior and inferior information are lost due to Placido disk configuration. Right map is generated from scanning slit data.

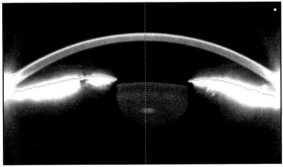

Figure 16-4. Example of a Pentacam Scheimpflug image of the anterior segment.

SCHEIMPFLUG TOMOGRAPHY

The Scheimpflug camera uses a laterally imaging light projected at 0° angle (Fig. 16-4).
1. The Pentacam (Oculus, Wetzlar, Germany) is a rotating camera that scans the anterior segment in 2 seconds taking 50 slits (up to 100 with the HR model.) It measures elevation and calculates curvature.

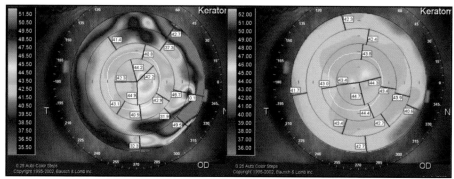

Figure 16-5. Instantaneous (left) and axial (right) maps of the same eye using the Keraton. Central curvatures are similar. Beyond 1.5 to 2 mm radial distance, instantaneous map shows flatter and more variable values, making the map noisier.

2. The Galilei (Ziemer, Port, Switzerland) has two rotating cameras scanning the anterior segment in 2 seconds taking 60 slits (Fig. 16-5). The advantage of the second camera is compensation of any corneal decentration through the image acquisition process. As does the Orbscan IIz, the Galilei has a built-in Placido disk. Corneal curvature is calculated by merging the Placido and elevation data by a proprietary algorithm.

Both Scanning Slit and Scheimpflug instruments measure anterior and posterior corneal curvature. This has the potential advantage of calculating central corneal power with paraxial or exact formulas instead of assuming posterior corneal curvature like Placido topographers do. Although most of the modern IOL calculation formulas have not been adapted to this possibility yet, some new formulas have been published in this way[4] and there is some software that calculates IOL power using exact ray tracing (Okulix[5]).

Curvature and Power Maps

Curvature can be expressed as axial radius of curvature, where the center of curvature of each point is forced to be at the optical axis. It can also be expressed as the instantaneous radius of curvature, where this constraint is not applied. The manual keratometer measures the axial radius of curvature at approximately a 1.6 mm radial distance.[6]

The difference between axial and instantaneous radius of curvature is not significant in the paraxial zone (Fig. 16-6C). Beyond the 2 mm radial distance, instantaneous radius of curvature describes corneal shape more accurately, giving a flatter reading in a normally prolate cornea.

From axial and instantaneous curvature (in mm), keratometric power (in D) is calculated by means of a paraxial refractive formula and a keratometric standard index of refraction of 1.3375 (see Chapter 14 for a full explanation). It is important to understand that this is a measure of shape and not of power. As a paraxial formula is used, a sphere will show the same power in the center as in the periphery (same color in the map), while due to Snell's law, power should increase in the periphery (hotter color than in the center) to express spherical aberration. Most videokeratoscopes calculate a refractive or Snell map where optical power of the anterior cornea is displayed.

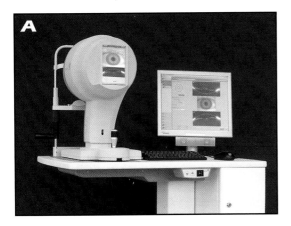

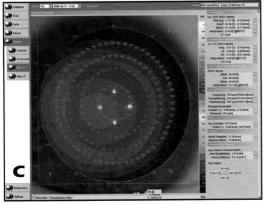
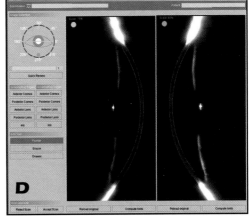

Figure 16-6. Ziemer Galilei instrument: (A) from examiner view, (B) from patient view, (C) elevation map, and (D) Scheimpfug photos of anterior segment.

Central Corneal Power/Radius

As modern IOL calculation formulas were designed to input keratometric K readings, most topographers calculate a simulated K measurement they call the SimK, which is calculated from the 3 mm ring analysis, disregarding more central curvature data. Some devices calculate other central indices, averaging central corneal curvature: Effective Refractive Power[7] (EyeSys), Average Corneal Power[8] (Tomey), etc.

Table 16-1.	
NORMAL EYES TOPOGRAPHIC DATA	
Author's database: SimK and asphericity measured with EyeTop (CSO) topographer; Anterior/posterior ratio measured with Pentacam.	
SimK	43.45 ±1.37
Asphericity (Q 4.5 mm)	-0.12 ±0.13
Asphericity (Q 8 mm)	-0.27 ±0.12
Anterior/Posterior	1.21 ±0.02

Accuracy and Precision

Placido videokeratoscopy has accuracy and precision within ±0.25 D measuring test spheres.[9] Douthwaite[10] found that the EyeSys overestimated apical radius 0.11 mm with ±0.02 mm 95% confidence limits when measuring ellipsoidal convex surfaces. Spherical surfaces were more accurately measured than aspherical. Repeatability of vertex radius was 0.014 mm. Lam et al[11] compared the Medmont E300 with a handheld autokeratometer reporting 0.24 D mean difference in K value with ±0.78 D and 95% limits of agreement. Potvin et al[12] reported inter-observer repeatability between 0.25 and 0.50 D both for TMS1 (Tomey) and EyeSys topographers. Giraldez et al[13] compared EyeSys, Nidek ARK700, and Javal and reported ±0.12 mm with 95% limits of agreement between EyeSys and Nidek and 0.17 mm between EyeSys and Javal. Dave et al[14] reported similar repeatability for EyeSys; ±0.072 D and for B+L manual keratometer, ±0.103 D. The 95% limits of agreement between both methods was ±0.34 D.

Topography for IOL Calculation

If the cornea is within normal limits of curvature, asphericity and anterior/posterior ratio (Table 16-1), manual or automated keratometry and videokeratoscopy will perform similarly in terms of accuracy and precision for IOL power calculation.[15]

As mentioned previously, 95% agreement limits show a wide range (>1 D in some studies) so different technologies can't be considered exchangeable: Bias must be calculated and corrected in the calculations (ie, through A constant adjustment.)

If the cornea has an abnormal curvature or asphericity, or an increased level of irregularity, corneal topography allows a more accurate evaluation of curvature than keratometry becoming probably more adequate for IOL power calculation. However, repeatability can be affected differently in each case depending on the type of irregular astigmatism. Thus, measuring repeated times with each method and averaging the K value is a reasonable approach. This is the case in corneal scarring, tear film irregularity, keratoconus, keratoplasty, etc.

If the anterior/posterior ratio is altered (as happens after corneal refractive surgery, where this ratio becomes a function of the induced change in anterior corneal shape) using a fixed corneal index of refraction to estimate the contribution of the unknown posterior surface is incorrect by definition. It leads to an overestimation of the SimK: 14% to

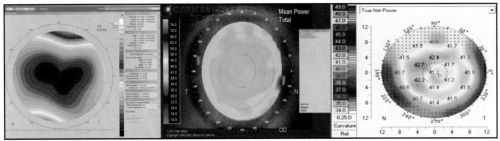

Figure 16-7. Keratometric power maps calculated from anterior and posterior radii. *Left:* Total Corneal Power by Galilei. *Center:* Mean total power by Orbscan. *Right:* True Net Power by Pentacam.

25% of the corrected diopters.[16,17] Many methods have been described to correct this fact and will be explained in a later chapter. A very simple way is to subtract (0.15*corrected diopters) from the topographic SimK.

The anterior/posterior ratio problem is overcome if both radii are accurately measured with corneal tomographers.

1. Srivannaboon et al[18] found a good correlation (r2=0.853) between surgery induced refractive change and Orbscan Total Optical Power for 4 mm with an average difference of +0.17 ±1.16 D.
2. Sónego-Krone et al[19] also reported a slight difference between refractive change and Orbscan Total Optical Power for 4 mm (+0.08 ±0.53 D) and Mean Total Power for 2 mm (+0.07 ±0.62 D).
3. Qazi et al[20] found the best correlation between the ideal SimK and the Orbscan Total Axial Power for 4 mm, Total Optical Power for 4 mm, and Mean Total Power for 3 mm with -0.16 ±0.64 D, +0.04 ±0.72 D and -0.24 ±0.68 D average differences respectively.

Pentacam calculates central corneal paraxial power from anterior and posterior radii in the True Net Power map. In order to use this value with the current IOL theoretical formulas (designed to input the SimK, calculated with n=1.3375) this value is converted to the Equivalent K Reading in the Holladay Report module of the Pentacam software. Galilei calculates similarly in the so-called Total Corneal Power map (Fig. 16-7).

New IOL calculation software will be marketed in the near future where real calculations, both paraxial and exact (performed with actual anterior and posterior corneal data) will avoid the need for the assumptions and conversions described in this chapter—making everything simpler (and hopefully more accurate).

References

1. Cohen KL, Tripoli NK, Holmgren DE, et al. Assessment of the power and height of radial aspheres reported by a computer-assisted keratoscope. *Am J Ophthalmol.* 1995;119(6):723–732.
2. Mattioli R, Carones F, Cantera E. New algorithms to improve the reconstruction of corneal geometry on the Keratron videokeratographer. *Invest Ophthalmol Vis Sci.* 1995;36(suppl):1400.
3. Mattioli R, Carones F. How accurately can corneal profiles heights be measured by placido-based videokeratography? *Invest Ophthalmol Vis Sci.* 1996;37(suppl):4273.

4. Norrby S. Using the haptic plane concept and thick-lens ray tracing to calculate intraocular lens power. *J Cataract Refract Surg.* 2004;30(5):1000–1005.
5. Preussner PR, Wahl J, Lado H. Ray tracing for intraocular lens calculation. *J Cataract Refract Surg.* 2002;28(8):1412–1419.
6. Douthwaite WA, Burek H. The Bausch and Lomb keratometer does not measure the tangential radius of curvature. *Ophthalmic Physiol Opt.* 1995;15(3):187–193.
7. Holladay JT. Corneal topography using the Holladay Diagnostic Summary. *J Cataract Refract Surg.* 1997;23(2):209–221.
8. Maeda N, Klyce SD, Smolek MK. Comparison of methods for detecting keratoconus using videokeratography. *Arch Ophthalmol.* 1995;113(7):870–874.
9. Hannush SB, Crawford SL, Waring GO, et al. Accuracy and precision of keratometry, photokeratoscopy, and corneal modeling on calibrated steel balls. *Arch Opthalmol.* 1989;107(8):1235–1239.
10. Douthwaite WA. EyeSys corneal topography measurements applied to calibrated ellipsoidal surfaces. *Br J Ophthalmol.* 1995;79(9):797–801.
11. Lam AKC, Chan R, Chiu R. Effect of posture and artificial tears on corneal power measurements with a handheld automated keratometer. *J Cataract Refract Surg.* 2004;30(3):645–652.
12. Potvin R, Fonn D, Sorbara L. In vivo comparison of corneal topography and keratometry systems. *Contact Lens Anterior Eye.* 1996;23(1):20–25.
13. Giraldez MJ, Yebra-Pimentel E, Parafita A, et al. Comparison of keratometric values of healthy eyes measured by Javal keratometer, Nidek autokeratometer and corneal analysis system (EyeSys). *ICLC.* 2000;27:33-38.
14. Dave T, Ruston D, Fowler C. Evaluation of the EyeSys model II computerized videokeratoscope. Part I: Clinical assessment. *Optom Vis Sci.* 1998;75(9):647–655.
15. Cuaycong MJ, Gay CA, Emery J, Haft EA, Koch DD. Comparison of the accuracy of computerized videokeratoscopy and keratometry for use in intraocular lens calculations. *J Cataract Refract Surg.* 1993;19(suppl):178–181.
16. Holladay JT. Cataract surgery in patients with previous keratorefractive surgery (RK, PRK and LASIK). *Ophthalmic Practice.* 1997;15:238–244.
17. Seitz B, Langenbucher A, Nguyen NX, Kus MM, Kuchle M. Underestimation of intraocular lens power for cataract surgery after myopic photorefractive keratectomy. *Ophthalmology.* 1999;106:693–702.
18. Srivannaboon S, Reinstein DZ, Sutton HCS, Holland SP. Accuracy of Orbscan total optical power maps in detecting refractive change after myopic laser in situ keratomileusis. *J Cataract Refract Surg.* 1999;25(12):1596–1599.
19. Sonego-Krone S, Lopez-Moreno G, Beaujon-Balbi OV, et al. A direct method to measure the power of the central cornea after myopic laser in situ keratomileusis. *Arch Ophthalmol.* 2004;122:159–166.
20. Qazi MA, Cua IY, Roberts CJ, Pepose JS. Determining corneal power using Orbscan II videokeratography for intraocular lens calculation after excimer laser surgery for myopia. *J Cataract Refract Surg.* 2007;33(1):21–30.

Corneal Power: Measuring Corneal Power With the Pentacam

Giacomo Savini, MD

The Pentacam (Oculus Inc, Wetzlar, Germany) is a rotating Scheimpflug camera that has been developed to image the anterior segment of the eye and has increased the range of available technologies for the estimation of corneal power. In contrast to videokeratography (VKG), where curvature data are derived from the measured distances between the rings projected onto the cornea, the Scheimpflug camera measures the corneal radius on the basis of acquired images of the cornea, via triangle calculation. This technology can provide the following corneal measurements: simulated keratometry (SimK), True Net Power, Equivalent K Reading (EKR), and Total Refractive Power; in addition it has been used to develop the BESSt formula.

Simulated Keratometry

This is calculated by entering the anterior corneal curvature radius (in meters (m)) into the thin lens formula for paraxial imagery, which considers the cornea a single refractive sphere and reads as:

$$\text{corneal power} = (n-1)/r$$

where r = corneal radius, n = standard keratometric index of refraction (the assumed value of the refractive index of the cornea and the aqueous humor combined, which is 1.3375 in the case of Pentacam), and 1.000 is the refractive index of air.

Although the mean SimK measured by the Pentacam does not show significant differences in comparison to standard VKG, the measurements of the two technologies cannot be used interchangeably as the agreement between the Pentacam and VKG is good, but not perfect: the 95% limits of agreement are approximately ±1.00 D. In other words, a difference of up to 1 D can be expected in 95% of eyes; such a range can cause relevant differences in the prediction of IOL power. Further study is required in order to determine which of the previously-mentioned instruments can assure the highest predictability in IOL power calculation.

True Net Power

This is the result of the Gaussian optics formula (GOF) for thick lenses, which reads as:

$$\text{corneal power} = (n_1-n_0)/r_1 + (n_2-n_1)/r_2 - [(d/n_1) \times (n_1-n_0)/r_1 \times (n_2-n_1)/r_2]$$

where n_0 = refractive index of air (= 1.000), n_1 = refractive index of the cornea (= 1.376), n_2 = refractive index of the aqueous humor (= 1.336), r_1 = radius of curvature of the anterior corneal surface (in meters), r_2 = radius of curvature of the posterior corneal surface (in meters) and d = corneal thickness (in meters).

In virgin corneas, the corneal power measured by this method is lower than SimK (about 1.2 D) and cannot be entered into 3rd generation formulas for IOL power calculation. Accordingly, different authors have already suggested correcting and re-optimizing the current formulas so that the corneal power calculated by GOF using the Pentacam data can be entered.[1]

Equivalent K Reading

This value, which is available in the Holladay report of the Pentacam (Fig. 17-1), is an adjustment of the True Net Power and aims to provide ophthalmologists with a corneal power measurement that can be used for IOL power calculation in eyes that have undergone excimer laser corneal refractive surgery. Measurements at the 1, 2, 3, 4, 4.5, 5, 6, and 7 mm zones are generated by the internal software. Even if the manufacturer suggests that the 4.5 mm measurement should be the preferred value, it has been reported[2,3] that the 2 and 3 mm measurements show the closest agreement with respect to the Clinical History Method. However, caution is warranted, as the 95% limits of agreement are still wide.

Total Refractive Power

The latest software version of the Pentacam (Version 1.17) provides an additional measurement, called Total Refractive Power. Corneal power is calculated according to Snell's law and using ray-tracing: parallel light is sent through the cornea and each light beam is refracted according to the different refractive index of air, cornea, and aqueous humor; the slope of the surfaces; and the exact location of the refraction. This measurement is promising, as it would not suffer from the keratometric index problem typical of eyes that have had excimer laser surgery, and awaits validation from clinical studies.

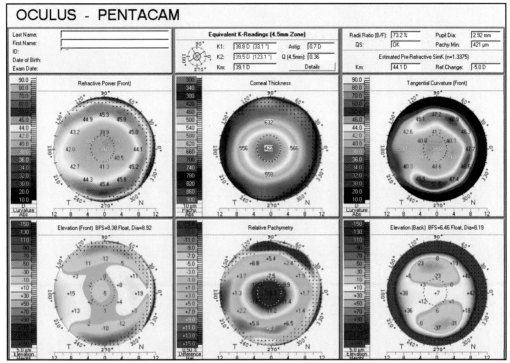

Figure 17-1. The Holladay report of the Pentacam of a patient that had undergone myopic PRK. The Equivalent K Reading at 4.5 mm is shown at the center top.

BESSt Formula

In order to achieve the best agreement with the Clinical History Method, Borasio[4] and coauthors modified the GOF and developed the BESSt formula, which is based on a variable rather than a static refractive index of the cornea. This formula has been implemented in a Microsoft Windows (Microsoft, Redmond, WA) software program and is also available in the freely downloadable Hoffer/Savini Tool from www.EyeLab.com. In preliminary studies, the values thus obtained lead to a good accuracy in IOL power calculation after myopic and hyperopic LASIK. Further studies, however, are necessary to confirm these results.

References

1. Norrby S. Pentacam keratometry and IOL power calculation. *J Cataract Refract Surg*. 2008;34(1):3.
2. Savini G, Barboni P, Profazio V, Zanini M, Hoffer KJ. Corneal power measurements with the Pentacam Scheimpflug camera after myopic excimer laser surgery. *J Cataract Refract Surg*. 2008;34(5):809–813.
3. Savini G, Barboni P, Carbonelli M, Hoffer KJ. Accuracy of Scheimpflug corneal power measurements for intraocular lens power calculation. *J Cataract Refract Surg*. 2009;35(7):1193–1197.
4. Borasio E, Stevens J, Sith GT. Estimation of true corneal power after keratorefractive surgery in eyes requiring cataract surgery: BESSt formula. *J Cataract Refract Surg*. 2006;32(12):2004–2014.

18

IOL Position: ACD and ELP

Kenneth J. Hoffer, MD

Axial Position of the IOL

The axial position for the IOL as it relates to the anterior surface of the cornea has historically been referred to as the *anterior chamber depth* (or ACD) because the optic of all IOLs in the early era was positioned in front of the iris, *in* the anterior chamber, or in the iris plane. Because almost all IOLs today are positioned behind the iris, new terminology has been offered such as *effective lens position* (ELP) by Holladay[1] and actual lens position (ALP) by the FDA. ELP is also used when referring to anterior chamber (AC) lenses, which do sit in the ACD.

Today, ACD is defined as the axial distance from the central front surface (anterior vertex) of the cornea to the central front surface of the crystalline lens. ELP is defined as the axial distance between the front surfaces of the two lenses (cornea and IOL); or, more exactly, the distance from the central front surface (anterior vertex) of the cornea to the effective principle plane of the IOL. For biconvex IOLs, the principle plane is located about in the middle of the IOL (50% of the thickness).

The ELP is required for all formulas, but is used in different forms. In the Binkhorst, Colenbrander, and Hoffer formulas, it is used directly and called the ACD. In the Hoffer Q formula, it is referred to as the pACD (personal ACD) and in the Holladay 1 formula it is calculated using a surgeon factor (SF) specific to each IOL style. In the SRK I and II and the SRK/T, it is incorporated into the A constant specific to each IOL style.

The formulas for converting these values are:

$$ACD = ((0.5663 * A) - 62.005)/0.9704$$

$$ACD = (SF + 3.595)/0.9704$$

$$A = (SF + 65.6)/0.5663$$

$$A = ((ACD * 0.9704) + 62.005)0.5663$$

$$SF = 0.9704 * ACD - 3.595$$

$$SF = 0.5663 * a - 65.6$$

where SF = surgeon factor, A = SRK A constant, and ACD = pACD.

These formulas result in rough approximations to begin with before developing individual personalized constants.

As an example, ACDs of 2.5, 4.0, and 5.5 mm would be converted to A constants of 113.78, 116.35, and 118.92, respectively. From these A constants you can see that the difference between an AC lens sitting at 2.5 mm (113.78) and a PC lens sitting at 5.5 mm (118.92) is about 5 D. Thus you can see that the ELP is contained within the A constant.

Some have proposed that it would be useful to measure the preoperative anatomic ACD (corneal epithelium to anterior capsule) either with an A-scan unit or by optical pachymetry. I performed such a comparison study[2] on 44 eyes and showed that the optical pachymetry method resulted in a mean 0.20 (±0.35) mm deeper ACD than obtained by ultrasound using 1548 m/sec (3.14 vs. 2.93 mm) and recommended the optical pachymetry for accurate measurement of ACD.

The IOL position has been considered the least important of the three variables as a cause of IOL power error, but in 1998 I examined a one-day postoperative (PO) IOL patient with a shallowed ACD and myopia of -2.50 D. After 3 days, the chamber deepened by 2.0 mm and the refractive error changed to plano. So it is found to be clinically important. IOL position has received the most attention from formula writers over the past 15 years. The major effort has been toward better prediction of where the IOL will ultimately rest.

In the 1970s and early 1980s, 3.5 mm was used as an ACD for all IOLs and all eyes. In 1982, I reported a study[3] of the postoperative position of PMMA posterior chamber (PC) lenses and noted that the PO ACD (or ELP) increased as the AL increased. From this data a formula was proposed to calculate the predicted ACD to use in the IOL power formula:

$$ACD = 2.92 * AL - 2.93$$

where AL = axial length.

Of course this regression formula was specific for this one lens style and would require adjustment for each IOL style. In 1988, Holladay referred to this AL-dependant method as the 2nd generation of formulas when he proposed the Holladay 1 formula as the 3rd generation, which introduced ACD dependence on the corneal curvature as well as the AL. He also introduced his lens-specific SF. He used a Fyodorov corneal height formula to calculate the distance from the front surface of the cornea to the iris plane and to that added the IOL-specific SF. The SF would be determined for each IOL style by back calculation knowing the PO results from a series of cases. This same process was used by Retzlaff when writing the SRK/T formula.

A study[4] I performed on a series of 270 eyes receiving a silicone plate haptic lens showed that the IOL shifted a mean of 0.06 mm posteriorly (ACD deepened) at 3 months PO, compared to its position on the first day after surgery. This correlated with a mean 0.21 D shift toward hyperopia seen in these patients.

The Olsen formula uses additional parameters, such as the preoperative ACD, corneal diameter, and lens thickness to predict the ELP. Holladay followed similar lines when devising the Holladay 2 formula. The Haigis formula dropped using the K and replaced it with the preoperative ACD.

NOTE: An IOL intended for capsular bag placement should be decreased by 0.75 to 1.25 D (depending upon the IOL power) when placed instead in the ciliary sulcus. Of course it would be wise to calculate a power for sulcus placement in advance by decreasing the ELP for that lens.

References

1. Holladay JT, Prager TC, Chandler TY, Musgrove KH, Lewis JW, Ruiz RS. A three-part system for refining intraocular lens power calculations. *J Cataract Refract Surg.* 1988;14(1):17–24.
2. Postoperative Measurement of Anterior Chamber Depths. Second U.S. Intraocular Lens Symposium, American Intra-Ocular Implant Society (AIOIS). Los Angeles, CA, April 1979.
3. Hoffer KJ. Biometry of the posterior capsule: A new formula for anterior chamber depth of posterior chamber lenses. In: Emery JC, Jacobsen AC, eds. *Current Concepts in Cataract Surgery: Selected Proceedings of the Eighth Biennial Cataract Surgical Congress.* Norwalk, CT: Appleton-Century Crofts; 1983;56–62.
4. PO Pachymetry ACD of 230 Staar AA4203 Silicone Plate IOL's. AAO Course #256, Modern Implant Surgery: XVI (Hoffer). American Academy of Ophthalmology; October 31, 1995; Atlanta, GA.

19

IOL Position: Measuring the ACD by Optical Pachymetry

Kenneth J. Hoffer, MD

Distances in the anterior chamber can be measured in a variety of ways. The earliest method was by optical pachymetry using a device developed by Goldman (Haag-Streit USA) for the slit lamp. I am most familiar with the device available for the Haag-Streit unit (Fig. 19-1). There is also one available for the Zeiss slit lamp (Fig. 19-2). I was first influenced to purchase this instrument after reading the studies by Cornelius Binkhorst,[1] who used it to measure the position of prepupillary IOLs in the supine and prone positions in the early 1970s.[1] It was quite a feat performing the examination in these two positions and his technician, Leo Loones, had quite a time. After receiving the unit, I taught myself how to use it and began using it on all my cataract patients.

Today, most ophthalmologists are unfamiliar with these methods since the introduction of and popularity of ultrasound pachymeters in the 1980s. The latter instruments tend to measure the depth of the anterior chamber 0.20 to 1.0 mm shorter than by optical pachymetry. I have long questioned the accuracy of measuring the anterior chamber depth (ACD) preoperatively or postoperatively using A-scan ultrasound, especially by contact applanation but even with immersion. Koranyi et al[2] compared contact applanation A-scan (using 1532 m/sec average sound velocity) with Scheimpflug imaging, a Haag-Streit optical pachymetry, and Orbscan analysis. Their results (Table 19-1) show the 26% to 28% increase in AC depth obtained by the optical modalities versus ultrasound.

These authors opined that the 1.0 mm shorter ACD may have been due to the indentation of the cornea in the contact method, but indentation of that magnitude by experienced technicians seems highly unlikely. Since the axial ACD is made up of the aqueous (at 1532 m/sec) and the cornea (at 1641 m/sec), the authors should have used an average

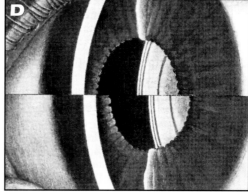

Figure 19-1. Optical pachymeters for measuring ACD using a slit lamp (A) Haag-Streit Optical Pachymeter II device (B) in its storage box with the split ocular, (C) viewing end of split ocular, (D) slit image as seen through split ocular.

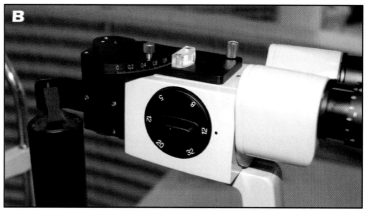

Figure 19-2. (A) Zeiss Optical Pachymeter in its storage box with the split ocular (B) attached to the top of the Zeiss slit lamp.

Table 19-1.

INCREASE IN AC DEPTH OBTAINED BY OPTICAL MODALITIES VS ULTRASOUND

Instrument	Pre-op ACD	PO ACD	Increase Over US	%
A scan US	3.05	3.69		
Scheimpflug	3.37	4.65	0.96	26%
H-S Optical		4.69	1.00	27%
Orbscan		4.71	1.02	28%

where US = ultrasound, H-S = Haag-Streit, Pre-op = preoperative, PO = postoperative

velocity of 1534 m/sec. However, if this discrepancy was due to using the wrong ultrasound velocity, the correct velocity would have had to be 1947 m/sec, which is not possible. I believe there is something inherent to the ultrasound measurement of the ACD that prevents getting an accurate reading by immersion or applanation. It appears that it would be more prudent to use other methods than ultrasound to measure the ACD, especially in scientific studies for publication.

The Zeiss IOLMaster also incorporates an optical measurement of ACD but I have found it to be quite variable compared to careful manual optical pachymetry using the Haag-Streit. It appears the optical pachymeter is presently the most accurate method to measure these AC distances.

Lastly, Suto and associates[3] performed a retrospective study of 30 normal eyes (22.0 to 24.5 mm) in which the same IOL (Alcon MA60BM) was implanted in the ciliary sulcus of one eye rather than the capsular bag due to capsular rupture. The physicians noted an average overcorrection in refractive error of -0.78 D at 3 months and a +1.11 D overcorrection in IOL power over and above the corresponding in-the-bag IOL. The difference between the mean measured ACD in these eyes (3.47 ±0.25 mm) and that of the contralateral eye (4.21 ±0.29) receiving the same lens in the bag was a statistically significant 0.74 mm (p<.0001). This translates into a 1.50 D/mm change in IOL position in normal AL range eyes. In order to lower the prediction error, the authors prospectively deducted 1.00 D from the planned IOL power when the lens was placed in the sulcus in 16 eyes. The prediction error then fell to -0.06 D ±0.19.

This chapter is intended to be a short course on how to take measurements with the Haag-Streit Optical Pachymeter II. The steps may seem complex and cumbersome but once one gets used to it, it is very easy to use.

Figure 19-3. Placing the pachymeter onto the Goldman tonometer post.

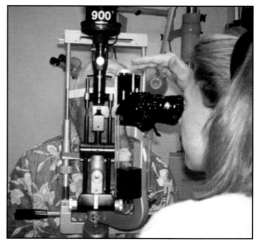

Figure 19-4. Slit lamp microscope housing turned 45° to the right of the slit beam housing.

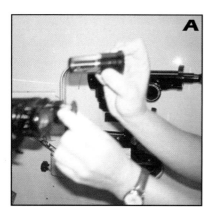
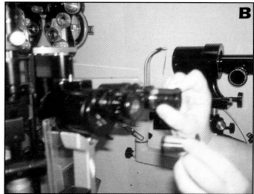

Figure 19-5. Replacement of the regular microscope right ocular (A) with the special split beam ocular (B).

Optical Pachymetry Examination

First remove the pachymeter from its storage box and slip it onto the Goldman tonometer post at the top of the slit lamp, making sure it is firmly seated (Fig. 19-3).

The slit beam housing should be aimed directly straight ahead at the patient's eye while you rotate the microscope housing 45° to the right of the slit beam housing. The examiner must now move the stool so that they can sit to the right side and comfortably take the reading (Fig. 19-4).

The regular microscope ocular for the right eye must now be removed (Fig. 19-5) and replaced with the special split beam ocular from the storage box (see Fig. 19-1).

Several adjustments have to be made. First, rotate the right ocular's magnification counterclockwise to +6 on the scale (Fig. 19-6).

Then you must rotate the split ocular so that the line in the opening is perfectly horizontal (Fig. 19-7).

Figure 19-6. Rotation of the right ocular's magnification to +6.

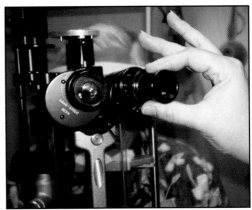

Figure 19-7. Rotation of the split ocular to horizontal.

Figure 19-8. Voltage switch; maximum is to the right.

Figure 19-9. Magnification turned to maximum.

Now the voltage switch of the slit beam, which is located at the bottom left of the slit lamp table, is turned to the right to its maximum level of illumination (Fig. 19-8).

Finally you must flip the objective magnification handle (below the binoculars) to maximum magnification (Fig. 19-9).

The patient's head can now be placed in the chin rest (Fig. 19-10) and the slit beam brought close to the patient and centered in the pupil by adjusting the slit using the other hand on the joystick.

Now the room door should be closed and all lights turned out so it is pitch-black. While looking through the right (split) ocular, use your right hand to rotate the gauge at the top of the instrument (Fig. 19-11).

You will see a single image (at reading zero) split into two (Fig. 19-12A) with one image sliding past the other as you rotate the gauge to the left (Fig. 19-12B).

Now rotate the gauge at the top of the pachymeter until the vertical slit image of the corneal front surface is perfectly aligned with the vertical slit image of the anterior lenticular surface (see Fig. 19-12B)—or whatever you desire to measure—making sure that the images are positioned in the center of the pupil. This latter requirement does take some constant

Figure 19-10. Patient placed in the slit lamp.

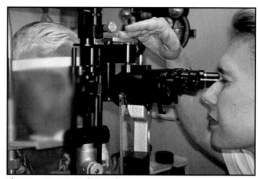

Figure 19-11. Taking the measurement.

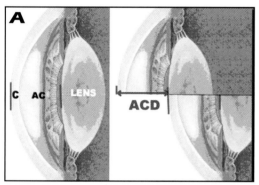

Figure 19-12. (A) Diagram of image of the anterior chamber through the split ocular; left at zero, right at ACD measurement. (B) Sliding the gauge.

adjusting of the slit lamp joystick to keep the image dead center. It helps to use a penlight and a hand magnifier to read the gauge scale and guesstimate to tenths of a millimeter.

The cornea magnifies or minimizes the image you are measuring, so the reading obtained must be corrected using the chart supplied by the manufacturer (Fig. 19-13). The chart uses radius of curvature rather than diopters, and the K readings and scale readings are so far apart that it requires a lot of mental interpolation on both scales.

To make this step easier I created a chart using a spreadsheet that first converted the K readings from radius of curvature to diopters and then expanded the ranges of both the X and Y axis to make the interpolation much less intense. The expansion was such that it required two charts which can be printed and laminated back to back (Fig. 19-14). The final reading is obtained by adding (the black numbers) or subtracting (the red numbers) the correction to the original pachymetry reading obtained. These charts can be obtained by requesting them by email (KHofferMD@AOL.com).

Once the process is learned, it becomes a very simple procedure to perform.

Figure 19-13. The Haag-Streit correction chart.

Anterior chamber depth Correcting table II
for anterior chamber depth measuring attachment II

Scale reading	Corneal radius / Correction (+ positiv - negativ)											
	5.5	6.0	6.5	7.0	7.5	8.0	8.5	9.0	9.5	10.0	10.5	11.0
0.20	0.01	0.01	0.01	0.01	0.01	0.01	0.01	0.01	0.01	0.01	0.01	0.01
0.40	0.02	0.02	0.02	0.02	0.02	0.02	0.02	0.02	0.02	0.02	0.02	0.02
0.60	0.01	0.02	0.02	0.02	0.02	0.03	0.03	0.03	0.03	0.03	0.03	0.03
0.80	0.01	0.01	0.02	0.02	0.02	0.03	0.03	0.03	0.03	0.03	0.04	0.04
1.00	0	0	0.01	0.02	0.02	0.02	0.03	0.03	0.03	0.04	0.04	0.04
1.20	0.02	0.01	0	0.01	0.01	0.02	0.03	0.03	0.03	0.04	0.04	0.04
1.40	0.03	0.02	0.01	0	0.01	0.01	0.02	0.03	0.03	0.04	0.04	0.05
1.60	0.05	0.04	0.02	0.01	0	0.01	0.02	0.03	0.03	0.04	0.04	0.05
1.80	0.07	0.05	0.04	0.02	0.01	0	0.01	0.02	0.03	0.04	0.05	0.05
2.00	0.09	0.07	0.05	0.03	0.02	0	0.01	0.02	0.03	0.04	0.05	0.06
2.20	0.11	0.09	0.06	0.04	0.02	0.01	0.01	0.02	0.03	0.04	0.06	0.07
2.40	0.13	0.10	0.07	0.05	0.03	0.01	0.01	0.02	0.04	0.05	0.06	0.07
2.60	0.15	0.12	0.09	0.06	0.03	0.01	0.01	0.03	0.04	0.06	0.07	0.09
2.80	0.17	0.13	0.09	0.06	0.04	0.01	0.01	0.03	0.05	0.07	0.09	0.10
3.00	0.18	0.14	0.10	0.07	0.04	0.01	0.02	0.04	0.06	0.08	0.10	0.12
3.20	0.19	0.15	0.11	0.07	0.03	0	0.03	0.05	0.08	0.10	0.12	0.14
3.40	0.21	0.15	0.11	0.06	0.03	0.01	0.04	0.07	0.10	0.12	0.15	0.17
3.60	0.21	0.16	0.10	0.06	0.02	0.02	0.06	0.09	0.12	0.15	0.17	0.20
3.80	0.22	0.15	0.10	0.05	0	0.04	0.08	0.12	0.15	0.18	0.21	0.24
4.00	0.22	0.15	0.09	0.03	0.02	0.06	0.11	0.15	0.18	0.22	0.25	0.28
4.20	0.22	0.14	0.08	0.02	0.04	0.09	0.14	0.18	0.22	0.26	0.29	0.32
4.40	0.21	0.13	0.06	0.01	0.07	0.12	0.17	0.22	0.26	0.30	0.34	0.38
4.60	0.20	0.11	0.04	0.03	0.10	0.16	0.21	0.26	0.31	0.36	0.40	0.44
4.80	0.18	0.09	0.01	0.07	0.14	0.20	0.26	0.32	0.37	0.42	0.46	0.50
5.00	0.16	0.07	0.02	0.10	0.18	0.25	0.31	0.37	0.43	0.48	0.53	0.57
5.20	0.14	0.04	0.06	0.15	0.23	0.30	0.37	0.43	0.49	0.55	0.60	0.65
5.40	0.11	0	0.10	0.19	0.28	0.36	0.43	0.50	0.57	0.63	0.68	0.74
5.60	0.08	0.04	0.15	0.25	0.34	0.42	0.50	0.57	0.64	0.71	0.77	0.83
5.80	0.05	0.08	0.19	0.30	0.40	0.49	0.57	0.65	0.73	0.80	0.86	0.92
6.00	0.01	0.12	0.25	0.36	0.47	0.56	0.65	0.74	0.82	0.89	0.96	1.02

Real value = Scale reading + Correction
Correction figures **Red** = negativ
Black = positiv
The table is based on an average refractive index of cornea and aequeous of 1.336

HAAG-STREIT INTERNATIONAL

References

1. Nordlohne ME. *The effect of supine and prone position of the patient on the position of the Binkhorst lens in the eye after intracapsular and extracapsular operation respectively.* The Hague, Netherlands: Dr. W. Junk B.V. Publishers; 1975:231-237.
2. Koranyi G, Lydahl E, Norrby S, Taube M. Anterior chamber depth measurement: A-scan versus optical methods. *J Cataract Refract Surg.* 2002;28(2):243–247.
3. Suto C, Hori S, Fukuyama E, Akura J. Adjusting intraocular lens power for sulcus fixation. *J Cataract Refract Surg.* 2003;29(10):1913–1917.

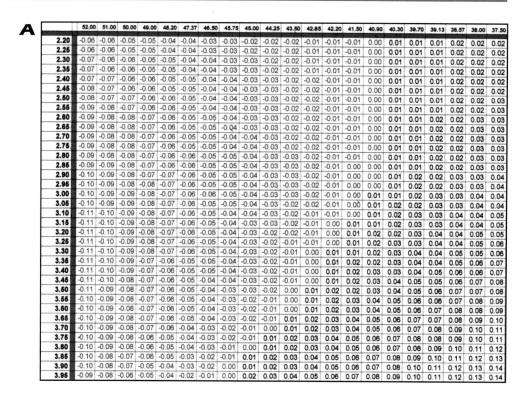

Figure 19-14. Correction charts to make the process easier.

IOL Position: Double-AL Method for Scleral Buckle Eyes

Kenneth J. Hoffer, MD

Modern 3rd generation formulas (except the Haigis) use both the AL and the K reading to predict the position the IOL will sit in the eye. When the central cornea is flattened by refractive laser surgery, the relationships in the anterior chamber do not change. However, now the formulas will be using a very flat K reading to predict the ELP. The Aramberri Double-K[1] method solves this problem by using two K readings; the preoperative K reading (or 43.5 D if unknown) to predict the ELP, and the postoperative flat K reading to calculate the IOL power. The Hoffer Programs® IOL power software allows the Double-K to be calculated for the Hoffer Q, the Holladay 1 and the SRK/T formulas and the Holladay IOL Consultant® allows it only for the Holladay 2 formula.

In the same vein, most post-encircling band retinal detachment (RD) eyes have an approximate 1.0 mm increase in AL, but the ACD is not affected by the encircling band (Fig. 20-1). Therefore, it would be best to use the measured AL minus 1 mm in the part of the formula that calculates the predicted ELP and use the measured AL in the part of the formula that calculates the IOL power; ie, Double-AL. This basically amounts to making the IOL power calculated a little weaker than would be predicted using the modern formulas. Since no IOL power programs automatically allow you to enter two ALs, alternatively one could just lower the power of the recommended IOL power in such RD eyes. I proposed this method in 2000.[2]

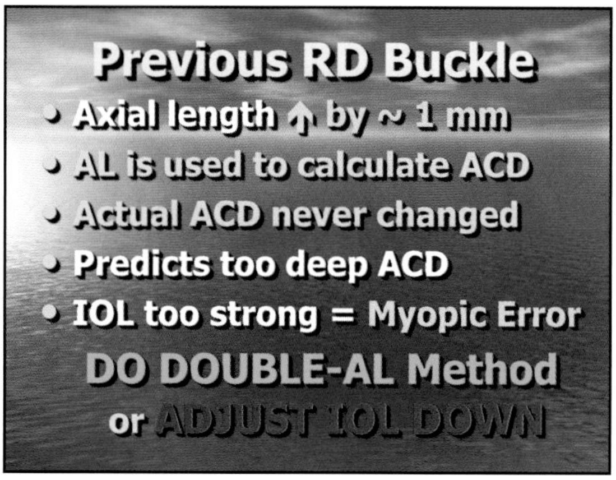

Figure 20-1. Rational for using a Double-AL in eyes with a scleral buckle.

> **Previous RD Buckle**
> - Axial length ↑ by ~ 1 mm
> - AL is used to calculate ACD
> - Actual ACD never changed
> - Predicts too deep ACD
> - IOL too strong = Myopic Error
>
> **DO DOUBLE-AL Method**
> or ADJUST IOL DOWN

References

1. Aramberri J. Intraocular lens power calculation after corneal refractive surgery: Double-K method. *J Cataract Refract Surg.* 2003;29(11):2063–2068.
2. Modernizing IOL Power. IOL Power Calculation: Striving for Accuracy. ASCRS Course #2102, Symposium on Cataract, IOL and Refractive Surgery. American Society of Cataract and Refractive Surgery; May 21, 2000; Boston, MA.

Formulas and Special Circumstances

Section II: Introduction

Kenneth J. Hoffer, MD

In the second half of this book we will be covering all the subjects dealing with IOL power calculation. It will be useful to you to have covered the first half of the book dealing with biometry, but it is not necessary. Here we will cover historical background on the formulas we use and the problems caused by the popularity of regression formulas. The newest theoretic formulas, including the Olsen and Haigis formulas will be covered; along with IOL power programs, formula personalization, and formula usage.

We will then cover most of the difficult problems that can face an ophthalmologist in trying to obtain an accurate IOL power, such as eyes with staphyloma (Fig. 21-1), eyes filled with silicone oil, unilateral high myopes and hyperopes, eyes also needing a penetrating keratoplasty, or eyes that have a scarred cornea.

Important today are the problems associated with eyes that have had refractive corneal surgery such as radial keratotomy (RK) as well as laser refractive surgery (PRK, LASIK and LASEK). We will cover the Double-K Method, problems with spherical aberration and how to calculate power for multifocal and toric IOLs as well as IOLs for pediatric eyes, piggyback IOL powers, anisometropia, and aniseikonia.

Finally we will give an overview as to how to prevent power errors and how to treat them when they occur. After going through all this material, it will become obvious that attention to detail can go a long way to providing the desired postoperative refractive error for a high percentage of your patients.

Figure 21-1. B-scan of an eye with staphyloma showing the bulging area above the optic nerve shadow.

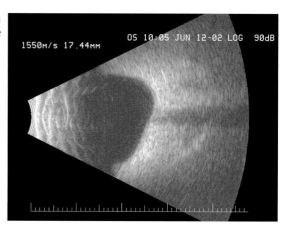

Formulas and Programs: Formula History and Basics

Kenneth J. Hoffer, MD

Formula History

It is important to understand the history of IOL power formulas so that one gains an understanding of the vagaries of today's modern formulas (Fig. 22-1).

1st Generation

The first IOL power formula was a theoretic one published by Fyodorov and Kolonko[1] in 1967. It was based on Gaussian optics and the Gullstrand eye. Gernet did extensive work in this regard in the late 1960s and 1970s also studying the effects of aneisikonia. Colenbrander[2] wrote his formula in 1972, followed by the Hoffer[3] formula in 1974 (published in 1981). Van der Heijde[4] published his formula and nomogram in 1975, the same year Binkhorst[5] published his formula. The latter became widely used in America in conjunction with the popularity of the Sonometrics 1st A-scan for IOL power. Because of dissatisfaction with the results of the Binkhorst formula, in 1978, first Lloyd and Gills,[6,7] followed by Retzlaff[7] and later Sanders and Kraff,[9] each developed a regression formula based on analysis of their previous IOL cases. Later Sanders, Retzlaff, and Kraff collaborated to produce the SRK I regression formula[10] in 1980. Since regression formulas were simple to calculate (P = A − 2.5 * AL − 0.9 * K) they were rapidly adopted as the standard throughout the world. The essence of this formula generation, theoretic or regression,

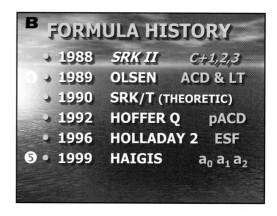

Figure 22-1. Formula history.

was that they all used a single constant for each lens that represented the predicted IOL position (ACD/ELP).

2nd Generation

The 2nd generation was issued by me[11,12] at the Welsh Cataract Congress in Houston in 1982. The results of a study of a large series of eyes showed a direct relationship between the AL of the eye and the position of the PMMA posterior chamber IOL. A simple regression formula was presented to better predict ACD:

$$ACD = 2.92 * AL - 2.93$$

Others followed by developing different mechanisms to apply this AL-related predictive relationship (Binkhorst,[13] SRK II[14]), which Holladay later defined as the second generation.

3rd Generation

In 1988, Holladay[15] proposed a direct relationship between the steepness of the cornea and the position of the IOL. He modified the Binkhorst formula to incorporate this as well as the AL relationship. Instead of ACD input, the formula would calculate the predicted distance from the cornea to the iris plane (using a corneal height formula by Fyodorov) and add to it the distance from the iris plane to the IOL. The latter he called the *surgeon factor* (SF) and it is specific to each lens. Since the SF is impossible to determine until after the IOL has been implanted, he calculated the SF for each lens style by using the PO refractive error, the IOL power implanted, along with the AL and K to back calculate through his formula what the ideal SF should have been. Then he would take the average of all the SFs in a large series to arrive at the SF to use for that lens in the future. The mathematics required a quadratic equation that necessitated a mathematician to solve.

In 1990, Retzlaff[16] followed suit and modified the Holladay 1 formula to allow use of an A constant instead of a SF, calling it the SRK/T theoretic formula. He did not personalize the A constant by back-calculating through the formula but instead relied on the older method of personalizing an A constant. The SRK/T was intended to replace the previous SRK regression formulas, but over the next decade 50% of American surgeons were still using the old regression formulas.

In 1992, at the urging of Holladay (personal communication), I developed the Q formula[17] using a tangent function of the K reading to accomplish a similar effect. The Fyodorov corneal height formula was **NOT** used. The base Hoffer formula (1974) was not changed. The Q part of the formula is just a separate calculation to predict the ELP used in the base formula. The Q formula starts with a personalized ACD (pACD) and adjusts it based on the AL and K. The personalized pACD is calculated for each lens style by back-calculating through the Hoffer Q formula, and the average of a series of ideal pACDs is used. This also required a quadratic equation and I also needed the aid of a mathematics professor to solve it for me.

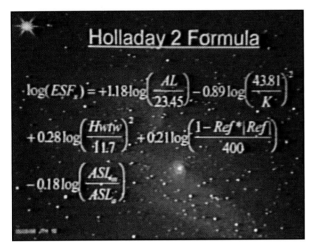

Figure 22-2. Holladay 2 formula.

4th Generation

In 1990, Olsen[18] proposed using other anterior segment measurements (such as the preoperative ACD) to better estimate the postoperative IOL position, and published algorithms for this. After several studies[19] showed the Holladay 1 formula not as accurate as the Hoffer Q in eyes shorter than 22 mm, Holladay, influenced by Olsen's concept, used the preoperative ACD measurement as well as the corneal diameter, the crystalline lens thickness, the preoperative refractive error, and age to calculate an *estimated scaling factor* (ESF). The IOL-specific ACD is then multiplied by the ESF to arrive at the ELP used in the formula. This he called the Holladay 2 formula (Fig. 22-2) which he has promulgated since 1996, but has yet to be published. Not being published, it is not possible for others to know if the formula has been tweaked since 1996.

Figure 22-3. Wolfgang Haigis, PhD.

5th Generation

In 1999, Wolfgang Haigis[20] (Fig. 22-3) proposed using three constants to predict the position of the IOL based on the characteristics of the eye and the IOL to be used. The formula replaces the use of the K reading with using the Olsen concept of using the preoperative ACD measurement. It calculates the predicted PO ELP by:

$$ELP = a_0 + a_1 * ACD + a_2 * AL$$

where ELP = predicted IOL position, a_0 = an IOL specific constant, a_1 = a lens specific constant to be effected by the measured preoperative ACD, a_2 = a lens specific constant to be effected by the measured preoperative axial length, ACD = the measured axial distance from the corneal apex to the front surface of the lens, and AL = axial length.

As in all formulas, the constants must be "optimized" (personalized) to each IOL style and surgeon. Single optimization only optimizes the a_0 and creates accuracy equal to the Hoffer Q and Holladay 1, but triple optimization of all three constants creates additional accuracy. The problem is that triple optimization requires a series of 500 to 1000 eyes of one lens style and the eyes in the series must statistically cover all axial lengths from very short to very long. This may be quite difficult to achieve for the average surgeon.

Refraction Formula

In 1993, Holladay[21] published a formula to calculate the power of an IOL to be implanted into an aphakic eye or an ametropic pseudophakic eye (piggyback IOL), or a phakic eye having a phakic IOL implanted. It does not need the AL but requires the corneal power, preoperative refractive error, and desired postoperative refractive error, as well as the vertex distance of both. I do not recommend its use in aphakic eyes because the vertex distance is difficult to measure accurately and, due to the high power of their refractive error, greater errors can result. It is, however, a good check against the AL formula calculation.

Summary

Today, it is pretty well accepted that regression formulas such as the SRK I and SRK II should not be depended upon for patient IOL power calculation except in emergency situations and only in eyes in the normal AL range. They are especially a problem in eyes that have had refractive corneal surgery and in these cases should **NEVER** be used.

References

1. Fyodorov SN, Kolonko AI. Estimation of optical power of the intraocular lens. *Vestnik Oftalmologic* (Moscow). 1967;4:27.
2. Colenbrander MC. Calculation of the power of an iris clip lens for distant vision. *Br J Ophthalmol.* 1973;57(10):735–740.
3. Hoffer KJ. Intraocular lens calculation: The problem of the short eye. *Ophthalmic Surg.* 1981;12(4):269–272.
4. Van der Heijde GL. A nomogram for calculating the power of the prepupillary lens in the aphakic eye. *Bibl Ophthalmol.* 1975;83:273-275.
5. Binkhorst RD. The optical design of intraocular lens implants. *Ophthalmic Surg.* 1975; 6(3):17–31.
6. Gills JP. Regression formula. *J Am Intraocul Implant Soc.* 1978;4(4):163–164.
7. Gills JP. Minimizing postoperative refractive error. *Contact and Intraocular Lens Med J.* 1980;6:56–59.
8. Retzlaff J. A new intraocular lens calculation formula. *J Am Intraocul Implant Soc.* 1980;6(2):148–152.
9. Sanders DR, Kraff MC. Improvement of intraocular lens power calculation using empirical data. *J Am Intraocul Implant Soc.* 1980;6(3):263–267.
10. Sanders D, Retzlaff J, Kraff M et al. Comparison of the accuracy of the Binkhorst, Colenbrander and SRK implant power prediction formulas. *J Am Intraocul Implant Soc.* 1981;7(4):337–340.
11. Hoffer KJ. Biometry of the posterior capsule. In: Emery JC, Jacobson AC, eds. *Current Concepts in Cataract Surgery* (Eighth Congress). New York, NY: Appleton-Century Crofts; 1983:56–62.
12. Hoffer KJ. The effect of axial length on posterior chamber lenses and posterior capsule position. *Current Concepts in Ophthalmic Surg.* 1984;1:20–22.
13. Binkhorst RD. Biometric A-scan ultrasonography and intraocular lens power calculation. In: Emery JE, ed. *Current Concepts in Cataract Surgery: Selected Proceedings of the Fifth Biennial Cataract Surgical Congress.* St. Louis, MO: Mosby CV; 1987:175–182.
14. Sanders DR, Retzlaff J, Kraff MC. Comparison of the SRK II formula and other second generation formulas. *J Cataract Refract Surg.* 1988;14(2):136–141.
15. Holladay JT, Prager TC, Chandler TY, et al. A three-part system for refining intraocular lens power calculations. *J Cataract Refract Surg.* 1988;14(1):17–24.
16. Retzlaff J, Sanders DR, Kraff MC. Development of the SRK/T intraocular lens implant power calculation formula. *J Cataract Refract Surg.* 1990;16(3):333–340. Erratum: 1990;16(4):528.
17. Hoffer KJ. The Hoffer Q formula: A comparison of theoretic and regression formulas. *J Cataract Refract Surg.* 1993;19(6):700–712. Errata: 1994;20(6):677 and 2007;33(1):2–3.
18. Olsen T, Oleson H, Thim K, Corydon L. Prediction of postoperative intraocular lens chamber depth. *J Cataract Refract Surg.* 1990;16(5):587–590.
19. Hoffer KJ. Clinical results using the Holladay 2 intraocular lens power formula. *J Cataract Refract Surg.* 2000;26(8):1233–1237.

20. Haigis W. The Haigis formula. In: HJ Shammas, ed. *Intaocular Lens Power Calculations.* Thorofare, NJ: SLACK Incorporated; 2003:41–57.
21. Holladay JT. Refractive power calculations for intraocular lenses in the phakic eye. *Am J Ophthalmol.* 1993;116(1):63–66.

Formulas and Programs: Regression and Theoretic Formulas

Kenneth J. Hoffer, MD

Accuracy Reporting

In reporting results of IOL power calculation or instrument performance it has been long established that the following data should be reported:

1. The mean error (ME) and standard deviation (SD) in prediction.
2. The mean absolute error (MAE) and standard deviation (SD) in prediction.
3. The percentage of eyes ±0.50 D from predicted target refraction.
4. The percentage of eyes ±1.00 D from predicted target refraction.
5. The percentage of eyes >2.00 D from predicted target refraction.
6. The range of errors from maximum plus to maximum minus.
   Eg: -0.06 D/0.46 D/67%/90%/1%/+1.76 to -1.01 D

It should be the comparison of the actual postoperative refractive error with that predicted by the power calculation and **NOT** the comparison of the IOL power implanted with the IOL power predicted by the calculation to produce the actual PO refractive error. The difference between these methods of reporting is a factor of 1.25/1. Thus, comparison of different studies using the two methods cannot be made easily.

This reporting schema allows one to see the clinical effects in one's practice from the results. Not only is the percentage of eyes within ±1.00 D important, but so is the

Table 23-1.

EYES WITHIN ±0.50 D OF PREDICTION DEPENDING ON AL OF THE EYE (HOFFER STUDY)

Best Results

Size	Formula	% ± 0.50 D	# Eyes
Short	Hoffer Q	67%	24
Medium	Hoffer Q	67%	219
Medium long	Holladay	71%	47
Very long	SRK/T	57%	13
			=303

303/450 = 67%

maximum range of errors as regards to IOL removals and medicolegal situations. It is important to remember that a study of 2 eyes, one with a -10 D error and the other with a +10 D error, results in a ME of 0.00 D and a MAE of 10.0 D.

Many studies on formula or instrument accuracy fail to report all the above information. They also report error in IOL power rather than in postoperative refractive error, and then convert the IOL power to a refractive error using an erroneous conversion factor not valid for all AL ranges. Still others use bilateral eyes in the same patient, which detracts from the statistical quality of results. All these discrepancies make comparing studies quite difficult.

Formula Accuracy

After the SRK[1] regression formula became very popular, there were reports of large errors clinically using the formula. I performed a small study comparing the SRK to the Hoffer (1974) theoretic formula[2] and noted a dramatic increase in errors using SRK. If one plots the results with any regression formula, it can be seen that it forms a straight line from the longest eye to the shortest. The same plot using a theoretic formula based on the actual optics of the eye show dramatic changes in the slope of the line as the eye becomes shorter or longer. This is the inherent problem with regression formulas in eyes outside the normal AL range of 22 to 24.5 mm.

However it wasn't until Holladay published the Holladay 1 formula[3] in 1988, that attention was paid to this problem. My large study[4] of 450 eyes (by one surgeon using one IOL style) compared the new Hoffer Q formula with the Holladay 1, the SRK/T[5], and the SRK I and II. The results (Table 23-1) showed that in the normal range (72% of the eyes) of AL (22.0 to 24.5 mm) almost all formulas function adequately, but that the SRK I formula is the leading cause of poor refractive results in eyes outside this range. The difference between the Hoffer Q and the SRK I and SRK II was statistically significant to a p value of 0.004 (Fig. 23-1).

The study also showed that the Holladay 1 formula was the most accurate in medium long eyes (24.5 to 26.0 mm)—which was 15% of the eyes—and the SRK/T was more accurate

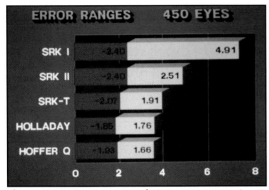

Figure 23-1. Error ranges of 450 eyes comparing 2 regression and 3 theoretic formulas.

Figure 23-2. Prediction errors within ±0.50 D error for the four modern formulas.

Table 23-2.

HOFFER RECOMMENDED FORMULA USAGE DEPENDING UPON THE AL OF THE EYE

Recommended Formula Usage

- Hoffer Q   <24.5 mm
- Holladay 1   24.5 to 26.0 mm
- SRK/T   >26 mm
- Holladay 2 and Haigis OK, but require more data collection
- Never SRK I or II

in very long eyes (>26.0 mm)—which was 5% of the eyes (Fig. 23-2). In short eyes (<22.0 mm) (8% of eyes) the Hoffer Q formula was most accurate, and this was confirmed (p>0.0001) in an additional large study by myself[6] (unpublished) of 830 short eyes (supplied by James Gills, MD) as well as in a multiple-surgeon study by Holladay. Holladay has postulated that the other formulas overestimate the shallowing of the effective lens position (ELP) in these very short eyes. The recommendations made since that study are shown in Table 23-2.

A more recent study[7] was performed on 317 eyes, again using one style of IOL and one surgeon. Since the Holladay 2 formula is not published, the study had to use the Holladay IOL Consultant® computer program to analyze the results with each of the formulas. We have recently learned that there were programming errors in the Holladay IOL Consultant® for the SRK/T formula. The results are displayed in Table 23-3. The study showed that the Holladay 2 formula equaled the Hoffer Q in short eyes (<22 mm) but was not superior to it (Fig. 23-3).

The Holladay 2 was not as accurate as the Holladay 1 or Hoffer Q in average AL eyes (22 to 24.5 mm). What was most disappointing is that the Holladay 1 was more accurate than the Holladay 2 in medium long eyes (24.5 to 26 mm). It is in this range that the Holladay 1 excels.

It appears that in attempting to improve the accuracy of the Holladay 1 formula in AL extremes, the addition of more biometric data input has improved the Holladay 2

Table 23-3.

RESULTS OF ACCURACY OF FOUR THEORETICAL FORMULAS ON 317 EYES USING THE HOLLADAY IOL CONSULTANT® FOR ANALYSIS

Formula	Mean Absolute Error						All 317 Eyes	
	Short <22.0	Normal 22.0–24.5	Med Long 24.5–26.0	Very Long >26.0	All Long <24.5	All Eyes	Max Error	>±2 D Error
Holladay 2	0.72	0.56	0.51	0.49	0.50	0.55	–1.60	0%
Holladay 1	0.85	0.42	0.37	0.56	0.43	0.43	–1.44	0%
Hoffer Q	0.72	0.43	0.47	0.58	0.50	0.45	–1.61	0%
SRK/T	0.83	0.46	0.35	0.44	0.36	0.44	–1.45	0%
Average	0.78	0.47	0.42	0.52	0.45	0.47		
Best	H-Q H-2	H-Q H-1	S/T H-1	S/T	S/T			

Shaded = recommended formulas[6]

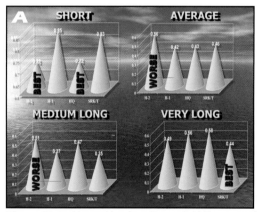
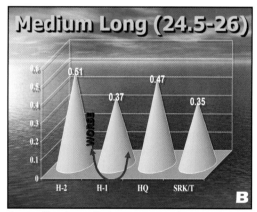

Figure 23-3. Mean Absolute Errors (MAE) comparing the Hoffer Q, Holladay 1 & 2, and SRK/T formulas. (A) Various AL ranges, and (B) medium long eyes showing the Holladay 2 having the worst MAE and the Holladay 1 the best.

formula in the extremes of AL, but deteriorated its excellent performance in the normal and medium long range of eyes (22.0 to 26.0 mm)—which is 82% of the population. It therefore does not appear to be necessary to collect all the data required for the Holladay 2 when the Hoffer Q, Holladay 1, and SRK/T only require AL and K readings.

On the other hand, there have been a few studies that report there is no formula AL dependence, and that the Hoffer Q, Holladay 1 and 2 and SRK/T formulas produce the same results regardless of the length of the eye.[8] There are also studies showing that the Hoffer Q formula is the most accurate in myopic eyes. However, most of these studies are

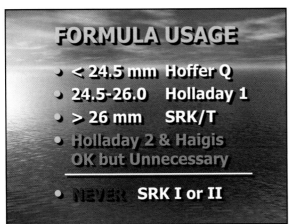

Figure 23-4. Recommended formula usage depending upon the AL of the eye.

based on aggregates of patients operated on by multiple surgeons, using a variety of IOLs and while mixing applanation, immersion and IOLMaster biometric measurements.

Finally, Holladay has stated that "every formula author is able to prove that his/her formula is the overall best" (personal communication). I broke that rule in 1993, by recommending the use of 3 formulas and proved that my own formula (Hoffer Q) was not the best in all eyes. Otherwise, since then, Holladay's dictum has held true. It is up to each ophthalmologist to critically look at the scientific data that has been reported and come to their own conclusion as to what will provide their patients with the greatest accuracy. A positive note is that the use of SRK regression formulas has decreased dramatically throughout the world.

Overall the recommendations for IOL formula usage based on AL still stand (Fig. 23-4) and have been statistically proven by a large 8000 eye UK study published in January 2011.[9]

References

1. Sanders D, Retzlaff J, Kraff M, et al. Comparison of the accuracy of the Binkhorst, Colenbrander and SRK implant power prediction formulas. *J Am Intraocul Implant Soc.* 1981;7(4):337–340.
2. Hoffer KJ. Intraocular lens calculation: The problem of the short eye. *Ophthalmic Surg.* 1981;12(4):269–272.
3. Holladay JT, Prager TC, Chandler TY, et al. A three-part system for refining intraocular lens power calculations. *J Cataract Refract Surg.* 1988;14(1):17–24.
4. Hoffer KJ. The Hoffer Q formula: A comparison of theoretic and regression formulas [published correction appears in: *J Cataract Refract Surg.* 1994;20(6):677 and *J Cataract Refract Surg.* 2007;33(1):2–3]. *J Cataract Refract Surg.* 1993;19(6):700–712.
5. Retzlaff J, Sanders DR, Kraff MC. Development of the SRK/T intraocular lens implant power calculation formula [published correction appears in: *J Cataract Refract Surg.* 1990;16(4):528]. *J Cataract Refract Surg.* 1990;16(3):333–340.
6. Accuracy of Hoffer Q Formula in 830 Short Eyes. 13th Biennial Welsh Cataract Congress, Cullen Eye Institute, Gallaria Hotel; September 9, 1994; Houston, TX.
7. Hoffer KJ. Clinical results using the Holladay 2 intraocular lens power formula. *J Cataract Refract Surg.* 2000;26(8):1233–1237.

8. Narvaez J, Zimmerman G, Stulting RD, Chang DH. Accuracy of intraocular lens power prediction using the Hoffer Q, Holladay 1, Holladay 2, and SRK/T formulas. *J Cataract Refract Surg.* 2006;32:2050–2053.
9. Aristodemou P, Cartwright NEK, Sparrow JM, Johnston RL. Formula choice: Hoffer Q, Holladay 1, or SRK/T and refractive outcomes in 8108 eyes after cataract surgery with biometry by partial coherence interferometry. *J Cataract Refract Surg.* 2011; 37:63–71.

Formulas and Programs: Olsen Formula

Thomas Olsen, MD

Significance of IOL Position Prediction

As described in Chapter 17, the position of the IOL after surgery is an important factor in the prediction of the refractive outcome of IOL implantation. If the IOL ends up deeper in the eye than expected, the effect is a hyperopic shift in refractive error. If it ends up more shallow, the effect is a myopic shift. The magnitude of the effect is such that about +0.7 mm shift in ACD corresponds to about +1.0 D shift in refraction with a strong dependence of the AL producing relatively higher errors in the short eyes.

Although it may be hidden to the user, all current IOL power calculation formulas contain their own method to arrive at an ELP (Table 24-1). At the time of the early theoretical formulas (eg, Colenbrander,[1] Hoffer,[2] Binkhorst[3]), very little was known regarding the actual position of the implant after surgery and a fixed value was therefore assumed. It soon became obvious that the fixed-ACD model was a bad assumption. The modern progress in IOL power calculation formulas has largely been made in the methods to predict the position of the implant after surgery based on preoperative measures.

Today there is indisputable evidence that the postoperative ACD or ELP is positively correlated with the axial length.[4] Therefore, to avoid a bias with the AL, the prediction of the postoperative ELP should in some way be corrected with the AL. The "way," however, is not easily defined, as it depends on the nature of the IOL calculation formula and the level of ambition.

Table 24-1.

VARIABLES USED TO PREDICT ELP BY VARIOUS FORMULAS

ACD/ELP Predictor	Formula/Author
AL	Hoffer[2] (original), Binkhorst II,[7] SRK/T,[9] Holladay 1,[10] Haigis,[11] Hoffer Q,[8] Olsen[6]
Tangent of K	Hoffer Q[8]
Corneal height	Fyodorov,[12] Holladay 1,[10] SRK/T,[9] Olsen,[6] Holladay 2*
Preoperative ACD	Haigis,[11] Olsen,[6] Holladay 2*
Lens thickness	Olsen,[6] Holladay 2*
Age	Olsen,[6] Holladay 2*
Refraction	Olsen,[6] Holladay 2*

*Not Published

If the IOL calculation formula is a "thin lens" formula (and most current formulas are), the IOL position is defined as the effective lens position (ELP), ie, the distance that correctly predicts the observed refractive effect when the corneal power and the AL are known. One way to proceed is to analyze a large series of cases to deduce the ELP and its dependence on the axial length, on the corneal power, and other preoperatively defined measures that might be considered.

If the IOL calculation formula is a "thick lens" formula (like the Olsen formula[5,6]), the IOL position is defined as the actual lens position (ALP), ie, the distance from the anterior central corneal surface to the anterior central surface of the IOL. This method requires a detailed knowledge of the optic configuration of the IOL in order to calculate the position of the anterior and the posterior principal planes of the IOL. The empirical methods to study the ALP dependence on preoperative measures do not differ from the "thin lens" approach, as it takes a large series to derive the exact dependency. This is often studied using statistical regression methods. It should be realized, however, that because of the differences between a "thin lens" and a "thick lens" model, the resulting regression coefficients are not interchangeable between the two approaches.

I believe that a "thick lens" approach gives a better and more realistic model, as it refers to the physical position and not the virtual position of the IOL implant. This may be a better path for the study of the anatomical dependence. More research seems required to resolve the issue.

ACD Models

CORNEAL CURVATURE AND CORNEAL HEIGHT

One of the earliest models for the prediction of the postoperative ACD was published by Fyodorov,[10] who used the base of the anterior spherical segment as the reference plane. He proposed that this plane could be calculated from the corneal curvature and the corneal diameter, the latter by taking an average value or by using the white-to-white measurement of the cornea. This Fyodorov formula was intended for iris-clip lenses and was adopted by myself for anterior chamber lenses[13] and later for posterior chamber lenses.[14]

The Fyodorov "corneal height" formula was reintroduced by Holladay[10] for the calculation of the so-called "surgeon factor", defined as the distance from the corneal height to the effective optical plane of the IOL and later adopted by the SRK/T formula.[9] Both of these formulas (Holladay 1 and SRK/T) were, in reality, a modified Binkhorst formula. However, recent work by this author seems to indicate that there is no significant information in the corneal height based on the corneal diameter, as compared with the corneal curvature itself. Other predictors like the AL, the preoperative ACD, and the lens thickness (LT) have been found to be more important.[15]

PREOPERATIVE ACD

Today, most of the newer generation IOL power calculation formulas recognize the importance of factors other than the AL to predict the ACD. One such predictor is the preoperative ACD which has been used in formulas such as the Haigis formula[11] and the Olsen formula.[5] From my studies, the importance of the preoperative ACD is ranked second to the AL in statistical significance as shown by multiple regression analysis.[15]

LENS THICKNESS

If one accepts the importance of the preoperative ACD in predicting the postoperative ACD, it seems logical to assume some influence of the preoperative LT as well. This is due to the thickening of the lens with age and the statistical negative correlation between ACD and LT in the normal eye.[20] Despite this logical assumption and the fact that most ultrasound equipment is capable of measuring the LT, it is surprising how little the LT has been used in ACD prediction algorithms. One exception to this rule is the Olsen formula (which has used this predictor since 1995[16]), and more recently it has also been considered by Norrby.[17] Recent studies on large series have confirmed that the LT is important for an accurate ACD prediction, especially in combination with the preoperative ACD.[5]

Comments

It should be noted that all studies to predict the postoperative ACD or the ELP from preoperatively defined measures require a normal anatomy of the eye. If this is not the case, such as eyes that have had keratorefractive surgery or if the AL is changed as a result of scleral buckling procedure (see Chapter 19), the statistical model behind the prediction of the ACD may no longer be valid and it may be necessary to "normalize" the anatomy. This is the rationale behind the Double-AL method of Hoffer and the Double K-method of Aramberri.[18]

Assuming the total prediction error in IOL power calculation to be the sum of the error associated with the main variables, namely measurement of AL, measurement of corneal power, and estimation of the postoperative ELP; it is possible to calculate the relative magnitude of each of these errors as previously shown by the author using ultrasound biometry.[19] The conclusion drawn from this study was that the AL constituted the largest source of error for the IOL power prediction, outranking the ACD error. This estimation is based on optimized IOL constants keeping the mean numerical error zero.

Assuming optical biometry is now the standard with its much higher reproducibility, the conclusion can be made that the statistical error arising from the estimation of the ELP is now the primary source of error (>40% of total statistical variance) among the total sources of error in IOL power calculation. The error from the measurement of the AL is second (>30%) and the corneal power error is third (remaining at 10% to 20%).[5]

Therefore, assuming optical biometry and optimized IOL constants, the accuracy by which the postoperative position of the IOL can be predicted is the major limiting factor of modern IOL power calculation formulas today.

References

1. Colenbrander MC. Calculation of the power of an iris clip lens for distant vision. *Br J Ophthalmol.* 1973;57(10):735–740.
2. Hoffer KJ. Intraocular lens calculation: the problem of the short eye. *Ophthalmic Surg.* 1981;12(4):269–272.
3. Binkhorst RD. The optical design of intraocular lens implants. *Ophthalmic Surg.* 1975; 6(3):17–31.
4. Hoffer KJ. Biometry of the posterior capsule. In: Emery JC, Jacobson AC, eds. *Current Concepts in Cataract Surgery* (Eighth Congress). New York, NY: Appleton-Century Crofts; 1983:56–62.
5. Olsen T. In: Shammas HJ, ed. *Intraocular Lens Calculations.* Thorofare, NJ: SLACK Incorporated; 2004:27–40.
6. Olsen T. Calculation of intraocular lens power: A review. *Acta Ophthalmol Scand.* 2007;85(5):472–485.
7. Binkhorst RD. Intraocular lens power calculation. *Int Ophthalmol Clin.* 1979;19(4):237–252.
8. Hoffer KJ. The Hoffer Q formula: A comparison of theoretic and regression formulas [published correction appears in: *J Cataract Refract Surg.* 1994;20(6):677 and *J Cataract Refract Surg.* 2007;33(1):2–3]. *J Cataract Refract Surg.* 1993;19(6):700–712.
9. Retzlaff J, Sanders DR, Kraff MC. Development of the SRK/T intraocular lens implant power calculation formula [published correction appears in: *J Cataract Refract Surg.* 1990;16(4):528]. *J Cataract Refract Surg.* 1990;16(3):333–340.
10. Holladay JT, Prager TC, Chandler TY, et al. A three-part system for refining intraocular lens power calculations. *J Cataract Refract Surg.* 1988;14(1):17–24.
11. Haigis W. The Haigis formula. In: Shammas HJ, ed. *Intraocular lens power calculations.* Thorofare, NJ: SLACK Incorporated. 2004:41–57.
12. Fyodorov SN, Galin MA, Linksz A. Calculation of the optical power of intraocular lenses. *Invest Ophthalmol.* 1975;14(8):625–628.
13. Olsen T. Prediction of intraocular lens position after cataract extraction. *J Cataract Refract Surg.* 1986;12(4):376–379.
14. Olsen T, Oleson H, Thim K, Corydon L. Prediction of postoperative intraocular lens chamber depth. *J Cataract Refract Surg.* 1990;16(5):587–590.
15. Olsen T. Prediction of the effective postoperative (intraocular lens) anterior chamber depth. *J Cataract Refract Surg.* 2006;32(3):419–424.
16. Olsen T, Corydon L, Gimbel H. Intraocular lens power calculation with an improved anterior chamber depth prediction algorithm. *J Cataract Refract Surg.* 1995;21(3):313–319.

17. Norrby S, Lydahl E, Koranyi G, Taube M. Clinical application of the lens haptic plane concept with transformed axial lengths. *J Cataract Refract Surg.* 2005;31(7):1338-1344.
18. Aramberri J. Intraocular lens power calculation after corneal refractive surgery: Double-K method. *J Cataract Refract Surg.* 2003;29(11):2063-2068.
19. Olsen T. Sources of error in intraocular lens power calculation. *J Cataract Refract Surg.* 1992;18(2):125-129.
20. Hoffer KJ. Axial dimension of the human cataractous lens [published correction appears in: *J Cataract Refract Surg.* 1993;111(12):1626]. *Arch Ophthalmol.* 1993;111(7):914-918.

Formulas and Programs: Accessing Modern IOL Power Formulas

Kenneth J. Hoffer, MD

To be able to use the modern theoretic IOL power formulas, it is easiest to use a computerized method, rather than doing it by hand as could be done with the outdated regression formulas. *There are several means by which the formulas may be used:*

1. Program them yourself from the published formulas.
   a. This is most fraught with difficulties because of the errata published for the Hoffer Q[1] and SRK/T[2] formulas, which are difficult to find and often ignored. It would be prudent to request the formula author to confirm that the programming has been done correctly prior to using them in a clinical situation.
   b. The author of the programming will be legally liable for any error in programming when used in a clinical setting.
   c. The Holladay 2 has never been published.
   d. Automatic personalization of IOL constants is more complicated to program.
   e. The Double-K method of formula calculation is more difficult to program.
2. Software installed on A-scan instruments.
   a. Be sure the formulas are duly licensed by their respective authors to insure that they have been programmed properly. The Hoffer Q and SRK/T formulas are the most prone to programming errors because of the published errata that are often ignored. Ask the manufacturer if the formula author has approved their programming of their formula.

b. One drawback is that personalization of IOL constants is usually not available and few offer database storage of patient data.
3. Software installed on the Zeiss IOLMaster.
   a. Formulas: Haigis,[3] Haigis-L, Hoffer Q, Holladay 1,[4] SRK II,[5] SRK/T.
   b. Be sure the **index of refraction in the Setup screen is set to 1.3375** when inputting external K readings to assure the Hoffer Q formula produces the proper result.
   c. Personalization of IOL constants is included along with database storage of patient data.
4. Software installed on the Haag-Streit LENSTAR LS900.
   a. Formulas: Hoffer Q, Holladay 1,[4] SRK II,[5] SRK/T.
   b. A database stores all the patient data.
5. Handheld programmable calculators.
   a. It is possible to program these units yourself, being careful to take into account the errata associated with the Hoffer Q and SRK/T formulas.
   b. Hoffer Programs® is available on a Casio calculator. It includes personalization of IOL constants.
6. Computer programs specific for IOL power calculation
   a. Hoffer Programs® is available for Windows and MAC.
      i. Formulas: Haigis, Hoffer Q, Holladay 1, SRK/T.
      ii. All IOL lens constants stored, possible to add new ones.
      iii. Automatic personalization of IOL constants.
      iv. Double-K available for Hoffer Q, Holladay 1 and SRK/T.
      v. Includes several methods to calculate K in LASIK eyes.
      vi. No annual maintenance fee.
   b. Holladay IOL Consultant® is available for Windows.
      i. Formulas: Hoffer Q, Holladay 1, Holladay 2, SRK/T.
      ii. All IOL lens constants stored, possible to add new ones.
      iii. Automatic personalization of IOL constants.
      iv. Double-K available *only* for Holladay 2.
      v. Excellent piggyback IOL calculation.
      vi. Annual maintenance fee involved.
7. Palm PDA programs.
   a. Hoffer Programs® is available for all Palm handhelds, including telephones running Palm OS Ver 6.0 (Fig. 25-1).
      i. Includes automatic personalization of IOL constants.
      ii. Includes clinical history and contact lens methods.
   b. DO NOT use free Palm program downloadable from a Russian website. All the formulas were checked by their respective authors and they are all programmed incorrectly.
8. iPhone, iTouch, iPad.
   a. Hoffer Programs® available at Apple app store.

Figure 25-1. Palm OS screens for Hoffer Programs®. (A) Data entry screen, (B) IOL calculation screen, (C) Formula personalization screen, (D) Resultant personalization factors.

History[6]

The first attempt to program IOL formulas was by Hermann Gernet in the early 1970s in Germany. He did this on a main-frame computer. I programmed my original Hoffer formula on the first programmable calculator, the Hewlett-Packard HP-65 in 1974; and later on a lineage of Casio programmable calculators. Later, Binkhorst had his formula programmed on a Texas Instruments calculator which came free with the purchase of the new Sonometrics A-scan unit. After publishing the Holladay 1 formula, Holladay made it commercially available on a Dioptron System calculator.

In 1988, I added the Holladay 1 to the Casio program, and later also added the SRK/T formula. This became commercially available as the "Hoffer Programs®" in 1990. The Hoffer formula was replaced with the Hoffer Q in 1993. In the same year, the first DOS and Windows computer program for IOL power was made available as Hoffer Programs®.

Table 25-1.

Formula Results Published in Original Article[7]
(Note Hoffer Q the worst.)

Formula	Mean	SD(D)	Range (D)
SRK II	+11.94	±7.07	+4.22 to +21.60
SRK/T	+4.40	±4.34	+0.40 to +11.17
Holladay 1	+2.74	±4.47	-0.56 to +10.20
Hoffer Q	+11.44	±7.49	+4.08 to +21.70

Table 25-2.

Formula Results Published in Journal Erratum[8]
Corrections by authors in bold. (Note Hoffer Q the best.)

Formula	Mean	SD(D)	Range (D)
SRK II	+11.94	±7.07	+4.22 to +21.60
SRK/T	+4.40	±4.34	+0.40 to +11.17
Holladay 1	**+3.03**	**±4.23**	-0.56 to +10.20
Hoffer Q	**+2.80**	**±1.83**	**-4.02 to +5.00**

This was followed several years later by the Holladay IOL Consultant® program available for Windows. Over the years various A-scan manufacturers added software to their ultrasound units to allow programming of various formulas. Several times they were programmed incorrectly, leading to clinical errors. This stimulated the formula authors to require licensing of the commercial use of their names so they could assure the software was programmed correctly.

Formula Propriety

The concept of formula propriety is often misunderstood. First, once it is published, a formula is free for anyone to use. It is not really possible to patent a formula. However, to be recognized as such, the formula usually has the name of its author. The author's name can be trademarked (™) or registered trademarked (®), thus requiring permission to use the name in commercial ventures. This allows the author to authenticate the formula programming and protect patients as well as the reputation of the formula and its author. There is no financial gain in licensing formulas as may be suspected. It is for the protection of the public.

The worst example of this problem was the major error by the Tomey ultrasound unit, which led to a published report by Oshika[7] and associates in 2001 showing that in a series of microphthalmic pediatric eyes, the Hoffer Q had a mean prediction error of 11.44 D compared with an error of only 2.74 D for the Holladay 1 and 4.40 D for the SRK/T (Table 25-1). The actual result, when programmed correctly, showed that the Hoffer Q formula was actually more accurate than the other 4 formulas in this series (Table 25-2), with a

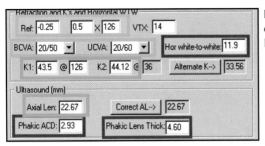

Figure 25-2. Holladay IOL Consultant® screen emphasizing the data required to calculate the Holladay 2 formula.

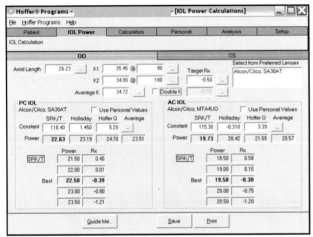

Figure 25-3. Hoffer Programs® screen showing calculation of the SRK/T formula in a myopic LASIK eye (23.51 D).

mean error of 2.78 D; a four-fold error of almost 9.0 D. The authors published an erratum[8] which states that "The error was caused by the incorrect power calculation program incorporated into the A-scan instrument (UD-7000, Tomey Corp.), which the authors failed to notice. Recalculation using the corrected program indicated that the Hoffer Q formula offered the most accurate predictions for the 5 microphthalmic eyes." It is noteworthy that their results for the Holladay 1 formula were also incorrect (see the Tables). Tomey immediately corrected their error in all their instruments and also issued an apology letter to ophthalmologists around the world. Licensing prevents such problems.

Popular Programs

The most popular commercial programs are the IOLMaster, the Hoffer Programs® System, and the Holladay IOL Consultant®, which include several formulas and the ability to personalize them as well as routines to deal with odd clinical situations. The Holladay IOL Consultant® is the only place the Holladay 2 formula is available. As illustrated in Fig. 25-2, the Holladay 2 requires the collection of other data than the AL and K reading (green boxes). The age of the patient, the preoperative refractive error (blue box), horizontal corneal diameter, phakic preoperative ACD, and phakic lens thickness (red boxes) must also be measured and entered to perform the calculation. The Aramberri Double-K method is available in the "Alternate K→" box, but this is not available for the Hoffer Q or SRK/T formulas.

The most beneficial development has been the availability of the Aramberri Double-K calculation for post-refractive eyes. Fig. 25-3 is an example of a Hoffer Programs® screen of a normal calculation for a myopic AL eye that has had LASIK, yielding a 23.51 D IOL

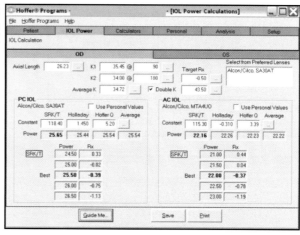

Figure 25-4. Hoffer Programs® screen showing calculation of the SRK/T formula in a myopic LASIK eye using the Double-K (25.54 D).

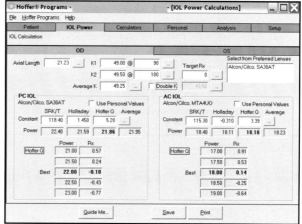

Figure 25-5. Hoffer Programs screen® showing calculation of the Hoffer Q formula in a hyperopic LASIK eye (21.95 D).

for emmetropia using the SRK/T formula. By clicking the "Double-K" box (Fig. 25-4) and entering the pre-LASIK K reading, the emmetropic power is now calculated as 25.54 D—thus preventing a 2 D error in IOL power. This would have resulted in more than 2.50 D of hyperopia.

A similar situation for a PO hyperopic LASIK eye is shown in Fig. 25-5. By clicking the "Double-K" box (Fig. 25-6) and entering the pre-LASIK K reading, the emmetropic power is now calculated as 20.59 D—thus preventing a 1.36 D error in IOL power. This would have resulted in 1.75 D of myopia.

References

1. Hoffer KJ. The Hoffer Q formula: A comparison of theoretic and regression formulas [published correction appears in: *J Cataract Refract Surg*. 1994;20(6):677 and *J Cataract Refract Surg*. 2007;33(1):2–3]. *J Cataract Refract Surg*. 1993;19(6):700–712.
2. Retzlaff J, Sanders DR, Kraff MC. Development of the SRK/T intraocular lens implant power calculation formula [published correction appears in: *J Cataract Refract Surg*. 1990;16(4):528]. *J Cataract Refract Surg*. 1990;16(3):333–340.

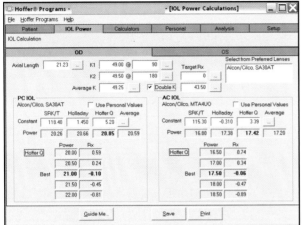

Figure 25-6. Hoffer Programs® screen showing calculation of the Hoffer Q formula in a hyperopic LASIK eye using the Double-K (20.59 D).

3. Haigis W. The Haigis formula. In: HJ Shammas, ed. *Intraocular Lens Power Calculations.* Thorofare, NJ: SLACK Incorporated. 2003:41–57.
4. Holladay JT, Prager TC, Chandler TY, et al. A three-part system for refining intraocular lens power calculations. *J Cataract Refract Surg.* 1988;14(1):17–24.
5. Sanders DR, Retzlaff J, Kraff MC. Comparison of the SRK II formula and the other second generation formulas. *J Cataract Refract Surg.* 1988;14(2):136–141.
6. Hoffer KJ. The history of IOL power calculation in North America. In: Kwitko ML, Kelman CD, eds. *The History of Modern Cataract Surgery.* The Hague, Netherlands: Kuglen Publications; 1998:193–208.
7. Oshika T, Imamura A, Amano S, et al. Piggyback foldable intraocular lens implantation in patients with microphthalmos. *J Cataract Refract Surg.* 2001;27(6):841–844.
8. Oshika T, Imamura A, Amano S, et al. Erratum. *J Cataract Refract Surg.* 2001;27:1536.

Formulas and Programs: Formula Personalization

Kenneth J. Hoffer, MD

The concept of personalizing a formula based on a surgeon's past experience and data was introduced by Retzlaff[1] after the introduction of regression formulas using an A constant in place of an anterior chamber depth (ACD). He proposed that the accuracy of regression formulas could be improved using this method. Here is how it works.

After an eye has had an IOL implanted, the following data is collected:
1. Preoperative axial length (AL)
2. Preoperative corneal power (K)
3. Implanted IOL power (P)
4. Stable postoperative refractive error (R)

Of the 5 parameters in all IOL power formulas, the only unknown factor now is the A constant. Using the regression formula, the A constant can be back-calculated. This is the ideal A constant that this eye would have needed to produce a perfect error-free IOL power prediction result. This can then be performed on a series of eyes and the average of all the ideal A constants can be calculated, producing the surgeon's ideal personalized A constant. But, as Retzlaff recommended, the series of eyes must be all the same, eg:
1. Same IOL model and manufacturer
2. Same surgeon
3. Same cataract procedure and IOL placement (eg, in or out of the bag)
4. Same AL measuring equipment
5. Same keratometer type

Eyes with postoperative surprises or acuities worse than 20/40 should not be included in the analysis, due to the poor accuracy in obtaining a precise refractive error. The minimum series size should be 20 eyes and, as more eyes are added, the benefit of the personalization increases.

This concept took on more importance in 1988 when Holladay[2] was writing his Holladay 1 formula. In attempting to better predict the postoperative IOL position (ELP), he used the Fyodorov corneal height formula to predict the distance from the anterior vertex of the cornea to the iris plane by using the AL and K. Since the IOL sat farther behind the iris plane, there was an additional distance that was missing; the distance from the iris plane to the principle plane of the IOL. He gave the value the name *surgeon factor* (SF). Since there was no way to calculate this number preoperatively, he back-calculated this value from a series of eyes previously operated on, took the average, and that became the SF to use on future cases. He also published a series of formulas to convert an ACD or A constant into a SF for a given lens style.

- When the SRK/T[3] theoretic formula was developed, they did not use this same method to personalize the new A constant to be used with it. Instead they used the old regression formula for personalization.
- When I[4] developed the Q formula for the Hoffer Q, I back-calculated for a personalized ACD and called it the pACD.
- When Haigis[5] developed his formula, he replaced corneal power with the preoperative ACD to predict ELP. He created three constants; a_0 (constant), a_1 (constant for AL), and a_2 (constant for ACD). Single optimization (personalization) of the Haigis formula leads to a personalized a_0 and its results are commensurate with the Hoffer Q, Holladay, and SRK/T formulas. Triple optimization results in personalized a_0, a_1, and a_2 and greater accuracy than the other formulas, but this requires a series of 500 to 1000 eyes of one lens style—which might be difficult for the average surgeon to acquire.

Several studies have proven that formula personalization definitely improves formula accuracy in a clinically significant way. Since it involves solving quadratic equations, the mathematics involved in the back-calculations is quite cumbersome. Both Holladay and I needed to resort to university math professors to accomplish it. Since computers will be needed, some instruments such as the IOLMaster will perform personalization. It has also been a prominent feature of both available IOL power calculation programs: Hoffer Programs® and Holladay IOL Consultant® (Fig. 26-1).

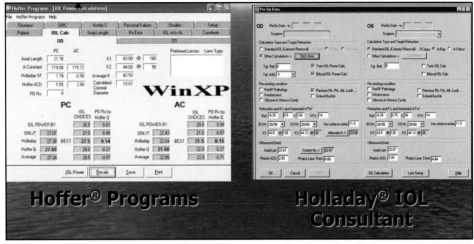

Figure 26-1. Personalization screen for Hoffer Programs® (left) and Holladay IOL Consultant® (right).

References

1. Retzlaff J. Calculating the surgeon's personal A-constant. In: Retzlaff J, Sanders DR, Kraff MC, eds. *Lens Implant Power Calculation Manual*, 3rd ed. Thorofare, NJ: SLACK Incorporated. 1990:12–13.
2. Holladay JT, Prager TC, Chandler TY, et al. A three-part system for refining intraocular lens power calculations. *J Cataract Refract Surg.* 1988;14(1):17–24.
3. Retzlaff J, Sanders DR, Kraff MC. Development of the SRK/T intraocular lens implant power calculation formula. *J Cataract Refract Surg.* 1990;16(3):333–340. Erratum: 1990;16(4):528.
4. Hoffer KJ. The Hoffer Q formula: a comparison of theoretic and regression formulas [published correction appears in: *J Cataract Refract Surg.* 1994;20(6):677 and *J Cataract Refract Surg.* 2007;33(1):2–3]. *J Cataract Refract Surg.* 1993;19(6):700–712.
5. Haigis W. The Haigis formula. In: HJ Shammas, ed. *Intaocular Lens Power Calculations*. Thorofare, NJ: SLACK Incorporated. 2003:41–57.

Special Circumstances: AL Measurement in Staphyloma Eyes

H. John Shammas, MD

The presence of a posterior pole staphyloma may be the most frequent condition in which a precise AL measurement may not be obtained. A posterior pole staphyloma is usually present in severe axial myopia, and the obliquity of the macular plane to the visual axis is responsible for significant AL differences within a small area. Occasionally it can be undetected if associated with a mature cataract in an eye with unilateral axial myopia.[1]

In A-scan biometry, the spatial orientation of the staphylomatous posterior pole surface causes an oblique rather than orthogonal interception of the ultrasound beam by the vitreoretinal interface.[2] This causes a saw-toothed aspect of the peak of the vitreoretinal interface (Fig. 27-1), which prevents precise localization of the foveolar area, and it becomes extremely difficult to select the right echogram (supposed to coincide with the visual axis). Measuring the AL with an IOLMaster can yield more accurate results, especially if the patient maintains good fixation.

In difficult cases, a B-mode guided biometry is the preferred technique, provided that the B-scan unit allows for such a measurement. A simplified immersion bath is created using the manually opened eyelid fissure filled with methylcellulose gel. The probe's tip is held in suspension within the gel layer, without touching the corneal surface. An optimal axial section of the eye is obtained and a control vector (seen on the screen as a superimposed dotted line) is aligned with the visual axis on the frozen image. An A-scan biometry is reconstructed along the control vector line (Fig. 27-2). It is important to visualize all echospikes corresponding to the anterior corneal surface, anterior and posterior lens surfaces, and the vitreoretinal interface in a location temporal to the optic nerve head. This technique is more popular in Europe and is routinely used in some centers.[3]

Figure 27-1. Immersion ultrasound measurement of a cataractous eye with posterior pole staphyloma. Note the low amplitude of the retinal spike, which in advanced cases might be difficult to identify.

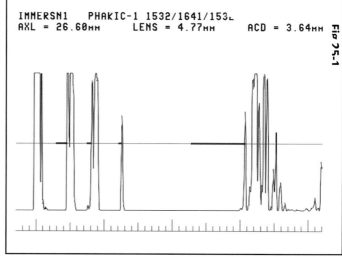

Figure 27-2. B-scan examination shows the increased curvature of the posterior pole in the cataractous eye with a central staphyloma. The A-scan is taken from the central vector axis. (Reprinted from O. Berges.)

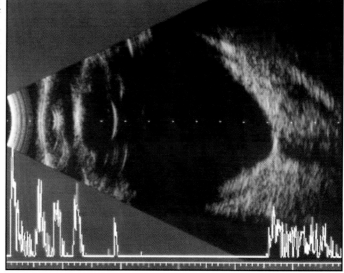

In addition, the B-scan offers a good evaluation of the vitreous cavity and of the retina, which is especially important in presence of a dense cataract and where the fundus cannot be evaluated.

References

1. Shammas HJ, Milkie CF. Mature cataracts in eyes with unilateral axial myopia. *J Cataract Refract Surg.* 1989;15(3):308–311.
2. Fernandez-Vigo J, Castro J, Diaz J, Cid MR. Ultrasonic forms of posterior staphyloma. *Ann Ophthalmol.* 1990;22(10):391–394.
3. Berges O, Siahmed K, Puech M, Perrenoud F. B-mode guided biometry. In: HJ Shammas, ed. *Intraocular Lens Power Calculations.* Thorofare, NJ: SLACK Incorporated. 2004:159–170.

Special Circumstances: Silicone Oil-Filled Eyes

Wolfgang Haigis, MS, PhD

Different media are characterized by different propagation velocities for ultrasound as well as light waves (eg, the speed of an ultrasound pulse through silicone oil is considerably slower—≈1/3—than through vitreous). Accordingly, it takes longer for the sound to cross the eye and it will return later than normal; just as if it had traveled a longer distance. This is why a normal eye with its vitreous replaced by silicone oil appears to be some 33 mm long when measured as a normal phakic eye. The necessary correction to the AL is nearly 10 mm.

The same physics apply to the propagation of laser light through silicone oil. The effect, however, is more than an order of magnitude smaller than with ultrasound. This is due to the differences in speeds of light for vitreous and silicone oil being only around 4%; the (group) refractive index of silicone oil is only slightly higher than that of vitreous. Compared to an IOLMaster measurement in phakic mode, a correction factor of some 0.7 mm has to be subtracted. This correction is automatically applied when the instrument's "AL Settings" for silicone oil-filled eyes are chosen.

There is a further difference between ultrasound and optical biometry: with ultrasound, segmental measurements can be performed—NOT so with the IOLMaster. Consequently, it is principally possible with ultrasound to determine and correct the vitreous distance in silicone oil for each eye individually, while the correction factor for IOLMaster biometry had to be derived for an average eye. Thus, a short eye may be falsely too short (≈50 μm) and a long eye may be falsely too long (≈100 μm). Clinically, these effects don't play a role.[1-3]

References

1. Parravano M, Oddone F, Sampalmieri M, Gazzaniga D. Reliability of the IOLMaster in axial length evaluation in silicone oil-filled eyes. *Eye.* 2007;21(7):909–911.
2. Habibabadi HF, Hashemi H, Jalali KH, Amini A, Esfahani, MR. Refractive outcome of silicone oil removal and intraocular lens implantation using laser interferometry. *Retina.* 2005;25(2):162–166.
3. Dietlein TS, Roessler G, Luke C, et al. Signal quality of biometry in silicone oil-filled eyes using partial coherence laser interferometry. *J Cataract Refract Surg.* 2005;31(5):1006–1010.

Special Circumstances: Unilateral High Myopes and Hyperopes

Kenneth J. Hoffer, MD

A vexing clinical problem faces the cataract surgeon when dealing with a relatively unilateral cataract in a patient with bilateral high ametropia. The dilemma is whether to make the surgical eye emmetropic or attempt to match the large ametropia of the other eye, which may never need surgery. If the surgical eye is made emmetropic (or relatively so), the resultant anisometropia will be intolerable. The only solution would be to wear a contact lens on the non-surgical eye. If the surgical eye is made refractively equivalent to the other eye, the patient misses the opportunity to be emmetropic for the first time in their life. This is a missed opportunity, especially if they do wind up developing a cataract in the 2nd eye. Of course it would be possible to perform a lens exchange or a piggyback IOL in the operated eye at a later date, but that would cause additional trauma to the eye.

I have always referred to the emmetropia option as "going for the brass ring." This option is possible if:
1. The patient is successful in a trial of monocular contact lens wear on the non-surgical eye.
2. The patient desires a phakic IOL in the other eye.
3. The patient desires a clear lens extraction on the other eye for refractive purposes.
4. The patient would prefer ignoring the ametropic eye and use the monocular vision of the newly emmetropic eye.

It has been my experience to convince most patients to accept a monocular CL, or ignore the other eye and go for the "brass ring" of emmetropia. Today it is also conceivable to place a piggyback lens over the emmetropic IOL to provide a refractive error that matches the non-surgical eye; this could easily be removed at a later date if the other eye ultimately needed surgery. Most studies have shown that the SRK/T formula[1] is more accurate for IOL power in extreme axial myopes and it is recommended that it be used for these eyes. In some cases a plano or negative-powered IOL may be required.

Haigis has recommended to me (personal communication) the use of special pACD for very low and negative IOL powers when using the Hoffer Q formula. Using an Alcon MA60, for example, he recommends a pACD of 15.94 for powers from +5.0 to 0.00 D and a pACD of -5.25 for powers from 0.00 to -5.0 D.

Since highly myopic eyes are more prone to retinal detachment following cataract surgery, it is wise to be more conservative in this regard and defray clear lens extraction in the non-cataractous eye. On the other hand, this is not the case with extreme hyperopes and it may be a better choice to perform clear lens extraction, which will result in a deeper anterior chamber and less chance for angle closure glaucoma. Most studies have shown that the Hoffer Q formula[2] and the Holladay 2 formula[3] are more accurate for IOL power in axial hyperopes (<22 mm) and it is recommended that it be used for these eyes.

Obtaining emmetropia in these eyes may require very high powers that are not commercially available in your favorite IOL style. It is wise to check with all IOL manufacturers to see if they have the power you need or are willing to special order it for you. If that is not possible, it will be necessary to piggyback two IOLs and attention must be paid to the calculations needed.

The anterior IOL will force the posterior IOL more posterior. This will decrease its effective power by moving its focal point behind the retina, causing hyperopia. The posterior movement is estimated to be about 50% of the thickness of the IOL, which can be calculated. Therefore, the total power needed has to be increased to make up for this and it should be split between the two IOLs, such that the majority of the power is in the posterior IOL. If there is an error, this will also make it easier to remove the thinner anterior IOL.

For some patients with milder ametropia, it may be advisable to plan a target refraction that is a compromise between emmetropia and the ametropia of the other eye such that anisometropia does not result.

In summary, it is extremely important to discuss these options with the patient and fully explain the advantages and disadvantages of each plan and allow them to make the decision as to which option to use. It is also prudent to make extensive notes of these conversations in the patient's chart.

References

1. Retzlaff J, Sanders DR, Kraff MC. Development of the SRK/T intraocular lens implant power calculation formula [published correction appears in: *J Cataract Refract Surg.* 1990;16(4):528]. *J Cataract Refract Surg.* 1990;16(3):333–340.
2. Hoffer KJ. The Hoffer Q formula: A comparison of theoretic and regression formulas [published correction appears in: *J Cataract Refract Surg.* 1994;20(6):677 and *J Cataract Refract Surg.* 2007;33(1):2–3]. *J Cataract Refract Surg.* 1993;19(6):700–712.
3. Holladay JT, Prager TC, Chandler TY, et al. A three-part system for refining intraocular lens power calculations. *J Cataract Refract Surg.* 1988;14(1):17–24.

30

Special Circumstances: Penetrating Keratoplasty and Scarred Corneas

Kenneth J. Hoffer, MD

Corneal Transplant Eyes

The problem with IOL power calculation in eyes scheduled for a combined procedure of cataract/IOL and penetrating keratoplasty (triple procedure) is attempting to predict what the corneal power will be after the corneal transplant. Some have suggested using either:

1. The corneal power of the other eye (if it is obtainable), or
2. Using an average of the surgeon's post-transplant corneal powers

The problem with these two options is that published reports show a very large range of prediction and refractive errors. This is especially unfortunate for patients that have already suffered enough from the poor vision due to corneal disease.

In 1986, I[1] published a suggestion that refractive results following a combined procedure would be better if the IOL implantation were performed as a secondary procedure after the corneal transplant has settled down. Thus one could either perform the corneal transplant alone and later perform the cataract/IOL surgery, or the corneal transplant can be combined with cataract removal followed by a secondary lens implant. In 1990, Geggel[2] reported excellent refractive results using this two-step approach. More than two-thirds of his eyes attained 20/40 or better uncorrected visual acuity. Ninety-five percent of the eyes were within ±2.00 D of the target postoperative refractive error (Fig. 30-1). What is most clinically significant, is that the total range of postoperative refractive error was only from +1.75 to -3.87 D instead of from +5.08 to -4.75 D if a triple procedure

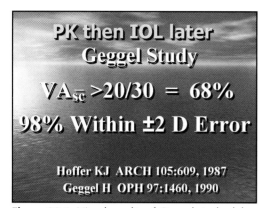

Figure 30-1. Visual results of Geggel study delaying lens implant until after PK.

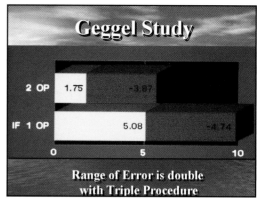

Figure 30-2. Range of error of Geggel study delaying lens implant until after PK.

had been performed using the surgeon's customary K reading selection. The total range of error dropped from 9.83 D to 5.62 D, a 57% decrease (Fig. 30-2). Though some may feel that two operations is a disadvantage, the refractive results are certainly an improvement for the patients.

An alternative to this concept would be to correct any residual ametropia after the triple procedure by implanting a secondary piggyback toric IOL or toric phakic IOL.

Corneal Scar Eyes

The problem of getting an accurate corneal power measurement in eyes with corneal scarring and irregular astigmatism has not received much attention. Cua et al[3] studied this in 2 eyes needing IOL exchange due to large postoperative IOL "surprises" of +5 D and -7.50 D each. They compared 6 methods to ascertain the corneal power and found the hard contact lens over refraction method to be the most accurate; decreasing the error they would have obtained with the manual keratometer of +4 to +5 D to -0.4 to -1.6 D. This may be a useful clinical option in such cases.

References

1. Hoffer KJ. Triple procedure for intraocular lens exchange. *Arch Ophthalmol.* 1987;105(5):609–610.
2. Geggel HS. Intraocular lens implantation after penetrating keratoplasty: Improved unaided visual acuity, astigmatism, and safety in patients with combined corneal disease and cataract. *Ophthalmol.* 1990;97(11):1460–1467.
3. Cua IY, Qazi MA, Lee SF, Pepose JS. Intraocular lens calculations in patients with corneal scarring and irregular astigmatism. *J Cataract Refract Surg.* 2003;29(7):1352–1357.

Special Circumstances: Radial Keratotomy Eyes

Giacomo Savini, MD

In 1985, 6 years after radial keratotomy (RK) was introduced in the United States, the problem of power calculation in RK eyes was first reported by Koch and associates,[1] and later by others.[2,3]

Correctly measuring the corneal power in eyes that have undergone RK is a difficult task due to the size of the optical zone (which is usually smaller than the area evaluated by keratometers or topography systems) so that measurements are likely to be performed in the region of the knee between the treated portion of the cornea and the flattened optical zone (Fig. 31-1). This may produce an overestimation of the corneal power, leading to an underestimation of the IOL power and resulting in unexpected postoperative hyperopia.

Inaccuracy of corneal power measurements may also be related, as in the case of excimer laser surgery, to the unreliability of the keratometric index (eg, 1.3375 or 1.3315) —which is known to assume a stable ratio between the anterior and posterior corneal curvature. Although several authors have previously stated that in incisional techniques like RK there is no loss of corneal tissue and both anterior and posterior corneal curvatures deform in parallel, so that the ratio between them is maintained. Actually there are no studies to confirm this statement. Using a Scheimpflug camera, we have recently observed that such a ratio is not maintained (data unpublished) and that the keratometric index should be adjusted in relation to the number of radial incisions.

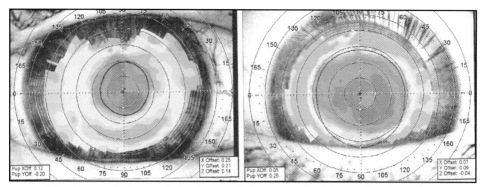

Figure 31-1. The optical zone following RK (left) is considerably smaller than after myopic LASIK (right).

It should also be observed that the unpredictability of IOL power calculation after RK can also be caused by the mechanical instability of the cornea following incisional surgery. Once phacoemulsification is performed, the RK incisions may temporarily reopen as if the RK incision procedure had just been carried out. This instability may temporarily exacerbate central flattening and peripheral bulging, which sometimes may persist.

However, the discrepancy between corneal topography measurements and true corneal power (as determined by the Clinical History Method[4,5] [CHM]) seems less than in the case of PRK and LASIK.

When phacoemulsification and IOL implantation are performed in eyes that have previously undergone RK, the choice of available methods to calculate the keratometric power (diopters) is considerably narrower with respect to the cases that received excimer laser surgery. If all preoperative data and the postoperative refraction are known, the CHM has been considered the standard for quite a long time; its reliability, however, may be limited by the frequent cases that experienced hyperopic shift after RK. For this reason, corneal topography probably represents a better option than CHM in these eyes. In a theoretical study by Stakheev and Balashevich,[6] the Sim-K generated by VKG seemed the most accurate among measured keratometric values. Other studies have also been reported.[7-10] Due to the small RK optical zones (usually <3.5 mm), it is now recommended to discard the Sim-K values and consider the more central area, such as the third ring or the Average Central Corneal Power (ACCP) of the TMS topographer (Tomey, Erlangen, Germany) or the Effective Refractive Power (EffRP) in the Holladay Diagnostic Summary[9] of the EyeSys Corneal Analysis System (EyeSys Vision, Houston, TX). A recent study by Awwad[11] and coauthors has shown that entering these values into Aramberri Double-K formulas leads to accurate results in IOL power calculation in these RK eyes.

References

1. Koch DD, Liu JF, Hyde LL, Rock RL, Emery JM. Refractive complications of cataract surgery after radial keratotomy. *Am J Ophthalmol.* 1989;108(6):676–682.
2. Cellikol L, Pavlopoulos G, Weinstein B, Cellikol G, Feldman ST. Calculation of intraocular lens power after radial keratotomy with computerized videokeratography. *Am J Ophthalmol.* 1995;120(6):739–750.

3. Chen L, Mannis MJ, Salz JJ, Garcia-Ferrer FJ, Ge J. Analysis of intraocular lens power calculation in post-radial keratotomy eyes. *J Cataract Refract Surg.* 2003;29(1):65–70.
4. Holladay JT. IOL calculations following radial keratotomy surgery. *Refract Corneal Surg.* 1989;5:36A.
5. Hoffer KJ. Calculation of intraocular lens power in post-radial keratotomy eyes. *Ophthalmic Practice* (Canada). 1994;12(5):242–243.
6. Stakheev AA, Balashevich LJ. Corneal power determination after previous corneal refractive surgery for intraocular lens calculation. *Cornea.* 2003;22(3):214–220.
7. Kim SH, Lee JH. Videokeratography to calculate intraocular lens power after radial keratotomy. *J Refract Surg.* 2004;20(3):284–286.
8. Packer M, Brown LK, Hoffman RS, Fine IH. Intraocular lens power calculation after incisional and thermal keratorefractive surgery. *J Cataract Refract Surg.* 2004;30(7):1430–1434.
9. Holladay JT. Corneal topography using the Holladay Diagnostic Summary. *J Cataract Refract Surg.* 1997;23(2):209–221.
10. Maeda N, Klyce SD, Smolek MK, McDonald MB. Disparity between keratometry-style readings and corneal power within the pupil after refractive surgery for myopia. *Cornea.* 1997;16(5):517–524.
11. Awwad ST, Dwarakanathan S, Bowman W et al. Intraocular lens power calculation after radial keratotomy: estimating the refracting corneal power. *J Cataract Refract Surg.* 2007;33(6):1045–1050.

Special Circumstances: Post-Laser Refractive Surgery Eyes

Kenneth J. Hoffer, MD

There are 3 main causes for the errors we see in IOL power calculation for eyes that have had corneal refractive surgery to correct ametropia.

1. **Instrument Error.** The problem of IOL power calculation errors in corneal refractive surgery eyes was first described by Koch et al[1] in 1989. The first problem that arises is that the instruments we use cannot accurately measure the corneal power needed in the IOL power formula in eyes that have had radial keratotomy (RK), photorefractive keratectomy (PRK), laser-assisted intrastromal keratomileusis (LASIK) and laser-assisted epithelial keratomileusis (LASEK). This major cause of error is due to the fact that most manual keratometers measure at the 3.2 mm zone of the central cornea, which often misses the central flatter zone of effective corneal power; the flatter the cornea, the larger the zone of measurement and the greater the error. The instruments usually overestimate the corneal power, leading to a hyperopic refractive error postoperatively.

2. **Index of Refraction Error.** The assumed index of refraction of the normal cornea is based on the relationship between the anterior and posterior corneal curvatures. This relationship is changed in PRK, LASIK, and LASEK but not in RK eyes. RK causes a relatively proportional equal flattening of both the front and back surface of the cornea, leaving the index of refraction relationship relatively the same. The other refractive procedures flatten the anterior surface but not the posterior surface thus changing the refractive index calculation, which creates an overestimation of

the corneal power by approximately 1 D for every 7 D of refractive surgery correction obtained. A manual keratometer measures only the front surface curvature of the cornea and converts the radius (r) of curvature obtained to diopters (D) using an index of refraction (IR) of usually 1.3375. The formula to change from diopters to radius is [r = 337.5/D] and from radius to diopters is [D = 337.5/r].

3. **Formula Error.** Most of the modern IOL power formulas (Hoffer Q,[2] Holladay 1,[3] and SRK/T[4]—but not the Haigis[5]) use the AL and corneal power (K) reading to predict the position of the IOL postoperatively. The flatter than normal K in RK, PRK, LASIK, and LASEK eyes causes an error in this prediction because the anterior chamber dimensions do not really change in these eyes.

History of Solutions

In 1989, Holladay[6] was the first to publish and popularize two methods to attempt to predict the true corneal power in refractive surgery eyes. I referred to them as the Clinical History Method and the Contact Lens Method.[7,8] The latter was first described by Frederick Ridley[9] in the United Kingdom in 1948 and introduced in the United States by Soper and Goffman[10] in 1974. Over the years many researchers and authors have proposed multiple methods to solve this problem. No one procedure has yet to be proven to be the most accurate in all cases.

In this regard Giacomo Savini of Bologna, Italy and I collaborated over a 2-year period to create an Excel spreadsheet tool that would automatically calculate most all the proposed methods and also provide a place to store all the data collected and entered. All the information could be stored in one place and it could be printed out on one sheet and stored in the patient's chart. The Hoffer/Savini LASIK IOL Power Tool (Fig. 32-1) was finished on July 4, 2007 and can be downloaded at no cost from www.EyeLab.com by clicking on the *IOL Power* button and then the *Hoffer/Savini* button.

In the creation of the Tool, we divided all the published methods into those that attempt to predict the true power of the cornea and those that fudge the target IOL power calculated with the standard data. We then divided each group into those methods that need historical data regarding the status of the patient's eye prior to refractive surgery and those that do not need any historical data.

Before finishing the Tool, we asked each formula author to beta test it to make sure they agreed with our calculations and assumptions. We have converted formula abbreviations to maintain consistency. The legend for these abbreviations is listed on Sheet #3 in the Tool and at the end of this chapter.

ASCRS has also developed a limited adaptation of this concept on their website at www.ascrs.org. Several methods are used to calculate an IOL power, but it has the shortcoming of not using the Hoffer Q or Holladay 2 formulas for short AL hyperopic LASIK eyes—which have been shown to be the most accurate in short eyes.

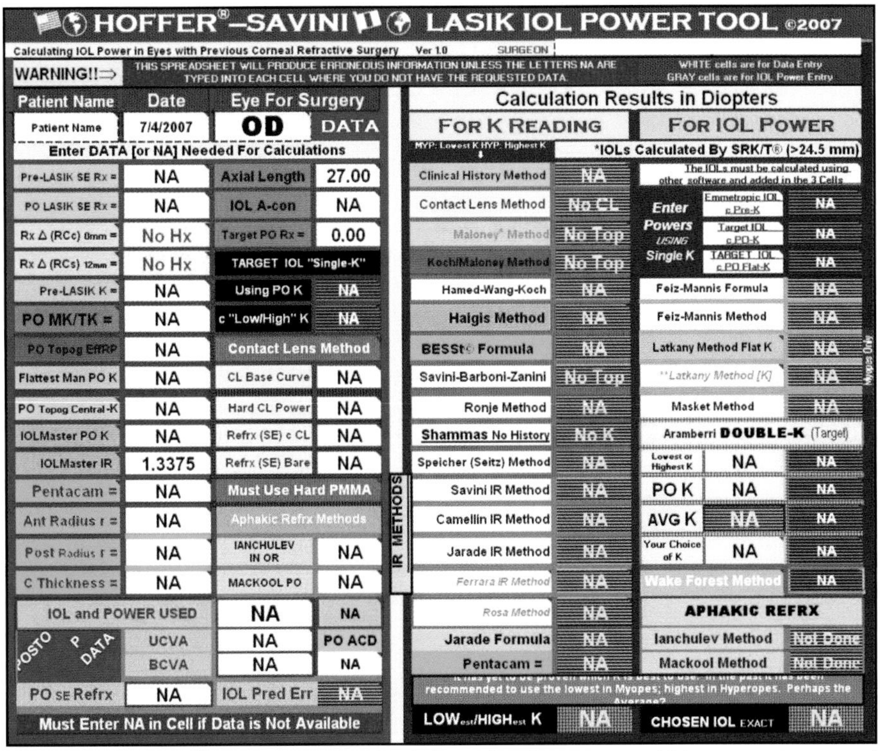

Special Circumstances: Post-Laser Refractive Surgery Eyes 181

Figure 32-1A. Hoffer/Savini LASIK Tool for refractive surgery IOL power (empty of data).

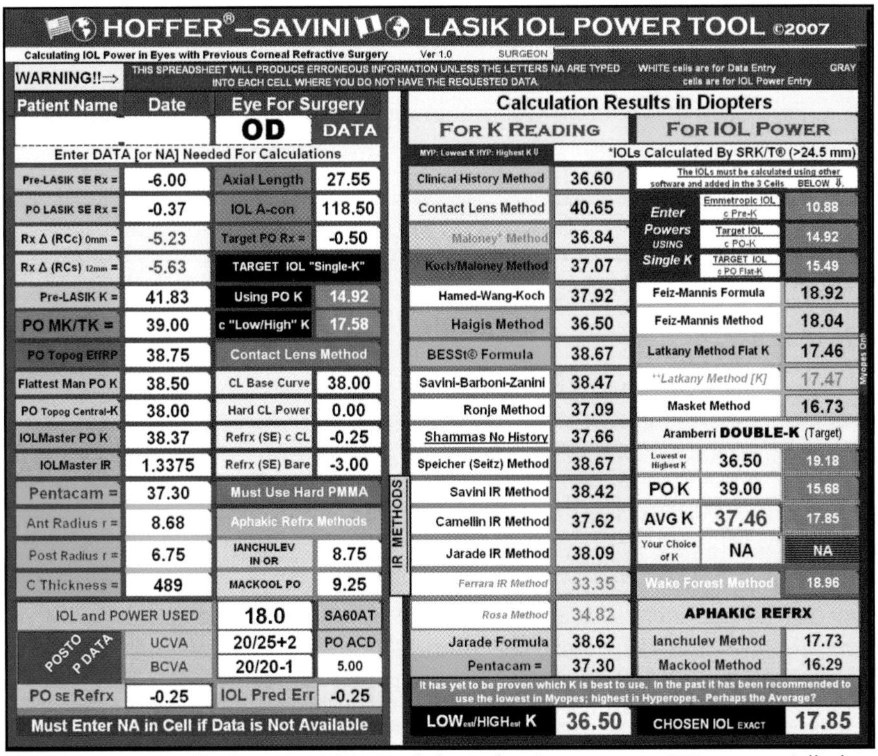

Figure 32-1B. Hoffer/Savini LASIK Tool for refractive surgery IOL power (all data entered).

Methods to Estimate True Postoperative Corneal Power

Those Needing Clinical History

Clinical History Method

$$K = K_{PRE} + R_{PRE} - R_{PO} \text{ or } [K = K_{PRE} + RC_C]$$

This method[1-9] is based on the fact that the final change in refractive error the eye obtains from surgery was due only to a change in the effective corneal power. If this refractive change the patient experienced is algebraically added to the presurgical corneal power, we will obtain the effective corneal power the eye has now. Obviously this requires knowledge of the K reading and refractive error prior to refractive surgery.

Originally it was recommended to vertex-correct the refractive errors to the corneal plane. Odenthal et al[11] showed that clinical results were better if they were not corrected. We have decided to use vertex correction in the Hoffer/Savini Tool because this is more scientifically accurate. Several IOL power calculation computer programs calculate the Clinical History method automatically when needed (eg, Hoffer Programs® and Holladay IOL Consultant®).

Hamed-Wang-Koch Method[12]

$$K = TK_{PO} - (0.15 * RC) - 0.05$$

This method requires knowledge of the refractive change from the surgery and the postoperative Sim-K from the topography unit.

Speicher[13] (Seitz[14,15]) Method

$$K = 1.114 * TK_{PO} - 0.114 * TK_{PRE}$$

This method requires obtaining the pre- and postoperative topographic Sim-Ks.

Jarade Formula[16]

$$K = TK_{PRE} - (0.376 * (TK_{POr} - TK_{PREr})/(TK_{POr} * TK_{PREr})$$

This method requires obtaining the pre- and postoperative topographic Sim-Ks in radius of curvature, not diopters.

Ronje Method[17]

$$K = K_{POFLAT} + 0.25 * RC$$

This method requires knowledge of the refractive change from the surgery and the postoperative flattest K reading measured now.

Adjusted Refractive Index Methods

These methods attempt to "correct" the index of refraction to better predict the corneal power. The first two methods require knowing the surgically induced refractive change at the spectacle plane and the average radius of curvature of the cornea now. The third method requires knowing the surgically induced refractive change at the corneal plane and the average radius of curvature of the cornea now.

1. Savini[18] Method: $K = ((1.338 + 0.0009856 * RC_S) - 1)/(K_{POr}/1000)$
2. Camellin[19] Method: $K = ((1.3319 + 0.00113 * RC_S) - 1)/(K_{POr}/1000)$
3. Jarade[20] Method: $K = ((1.3375 + 0.0014 * RC_C) - 1)/(K_{POr}/1000)$

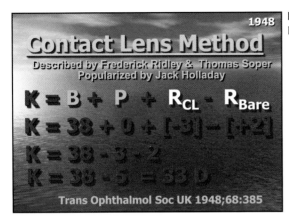

Figure 32-2. Example calculation of the Contact Lens Method.

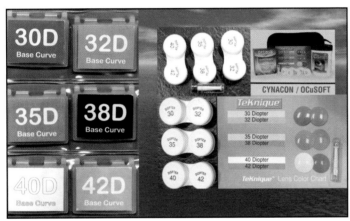

Figure 32-3. Rigid PMMA plano contact lens kit.

THOSE NOT NEEDING CLINICAL HISTORY

Contact Lens Method[9,10]

$$K = B_{CL} + P_{CL} + R_{CL} - R_{NoCL}$$

The Contact Lens Method was first described by Frederick Ridley[9] of England (the inventor of NaOH IOL sterilization) in 1948 and taught by Joseph Soper[10] in 1974. This method is based on the principle that if a hard PMMA (not rigid gas permeable) contact lens (CL) of plano power (P_{CL}) and a base curve (B_{CL}) equal to the effective power of the cornea is placed on the eye it will not change the refractive error of the eye. Therefore, the difference between the manifest refraction with the contact lens (R_{CL}) and without it (R_{NoCL}) is zero. The formula above computes the effective corneal power if there is a difference in any of these parameters (Fig. 32-2).

Originally it was recommended to vertex-correct the refractive errors to the corneal plane. Odenthal et al[11] showed that clinical results were better if they were not corrected. We have decided to use vertex correction in the Hoffer/Savini Tool because this is more scientifically accurate. Several IOL power calculation computer programs calculate this method and the Clinical History Method automatically when needed (eg, Hoffer Programs® and Holladay IOL Consultant®). Plano contact lens sets for performing this procedure are commercially available (Fig. 32-3).

Figure 32-4. Humphrey Topography axial map.

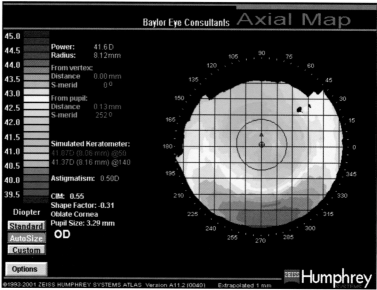

Obviously, this method is impossible if the cataract precludes performing an accurate refraction whereby the visual acuity is worse than 20/80.

Maloney Central Topography Method[21]

$$K = 1.1141 * TK_{PO\text{-}CTR} - 5.5$$

Based on his analysis of post-LASIK corneal topography (Fig. 32-4) central Ks (TK) on LASIK eyes, Maloney developed a formulation to predict true corneal power using *only the single central* postoperative reading TK.

Koch/Wang Method[22]

$$K = 1.1141 * TK_{PO} - 6.1$$

Koch and Wang analyzed several of these methods and obtained the best results using the Maloney method (discussed earlier) but only after increasing the constant from 5.5 to 6.1. They also offered a second method to calculate true corneal power if the change in the patient's refractive error (RC) is known. The formula is:

$$K = Kt_{PO} - (0.19 \times RC)$$

Savini-Barboni-Zanini Method[23]

$$K = 1.114 * Kt_{PO} - 4.98$$

This method only requires the postoperative Sim-K from topography.

Table 32-1.

ROSA CORRECTION FACTOR TABLE

AL range	RCF
22 to <23	1.01
23 to <24	1.05
24 to <25	1.04
25 to <26	1.06
26 to <27	1.09
27 to <28	1.12
28 to <29	1.15
>29	1.22

Shammas No History Method[24]

$$K = 1.14 * K_{PO} - 6.8$$

Shammas studied a series of eyes that had had LASIK. His analysis led him to propose a formula to predict the effective power of the cornea without needing any of the patient's clinical history, only the postoperative K reading obtained with manual keratometry.

Adjusted Refractive Index Methods

1. *Ferrara Method*[25]

$$K = ((-0.0006 * AL^2 + 0.0213 * AL + 1.1572) - 1)/(K_{POr}/1000)$$

This method requires the AL measurement and the postoperative K reading in radius of curvature.

2. *Rosa Method*[26]

$$K = (1.3375 - 1)/((K_{POr} * RCF)/1000)$$

This method requires the postoperative K reading in radius of curvature and the use of a table to obtain a factor (RCF) based on AL (Table 32-1). Unfortunately, they used the SRK II regression formula in their computation, which I disagree with.

3. *Haigis Method*[27]

$$K = -5.1625 * K_r + 82.2603 - 0.35$$

This method requires only the postoperative K reading form the Zeiss IOLMaster in radius of curvature (or converted to diopters using the index of refraction setting in the IOLMaster).

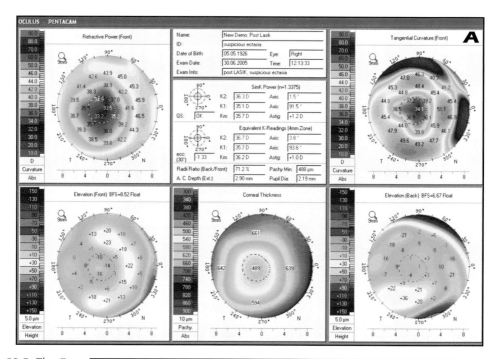

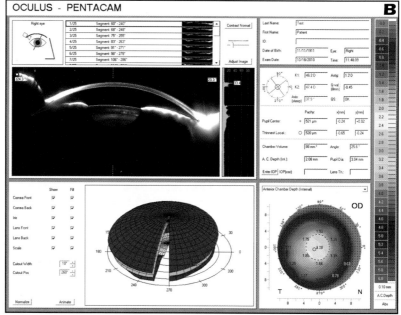

Figure 32-5. The Oculus Pentacam provides a topographic analysis of the corneal front and back surfaces as well as central corneal thickness.

Oculus Pentacam

A comprehensive Eye Scanner, the Oculus Pentacam (Oculus, Inc, Wetzlar, Gemany, www.oculususa.com) images the anterior segment of the eye by a rotating Scheimpflug camera measurement (Fig. 32-5). This rotating process supplies pictures in three dimen-

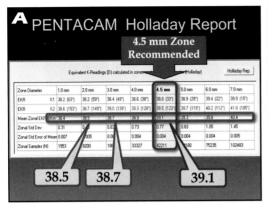
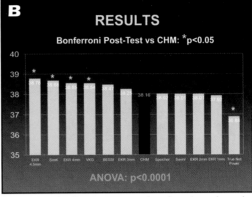
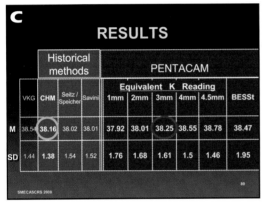

Figure 32-6. Pentacam study showing that the Holladay EKR recommendation of a 4.5 mm zone is less accurate than using a 3 mm zone compared to the Clinical History Method.

sions, provides a topographic analysis of the corneal front and back surfaces as well as central corneal thickness, and generates a TrueNetPower map of the cornea.

The TrueNetPower map of the postoperative cornea produced has been proposed as an accurate measure of the true corneal power. Initial results were disappointing and the software was reconfigured in early 2007. Fig. 32-6 shows the results reported by Savini et al[28-29] that demonstrate that the closest to the Clinical History Method is the Equivalent K from the Pentacam 3.0 mm zone, not the 4.5 mm zone as recommended. A UCLA study[30] has shown the Pentacam readings as much as 2.00 D off from the back-calculated K reading in post-LASIK eyes. There are several other studies on this new software that have also not lived up to expectations and newer changes are being proposed.

The BESSt Formula[31]

Published by Borasio, it uses the anterior and posterior corneal curvatures as well as the central pachymetry from the Pentacam unit to produce a predicted central corneal power. The formula is quite complicated, but it is incorporated into the Hoffer/Savini LASIK Tool. A Version 2.0 of the formula has just been released.

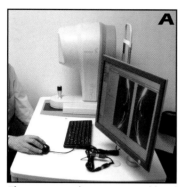
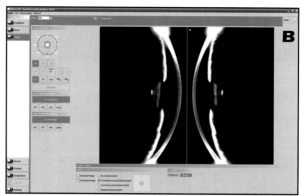

Figure 32-7. The Ziemer Galilei dual Scheimpflug camera. (A) Instrument, (B) anterior segment scans.

Figure 32-8. Galilei elevation maps.

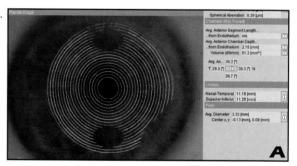
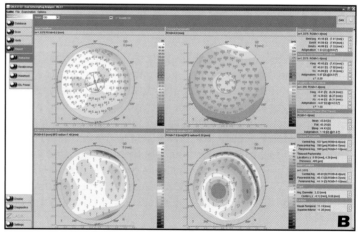

Ziemer Galilei

A new instrument has been introduced called the Galilei (Figs. 32-7 and 8) that utilizes two Scheimpflug cameras with a placido disk to attempt to better evaluate the anterior segment structures to produce a true corneal power. Studies are quite promising and ongoing.

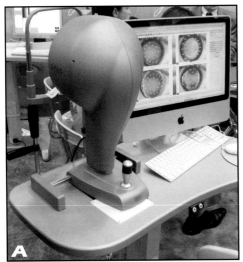
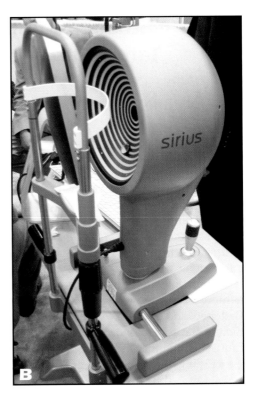

Figure 32-9. CSO Sirius Scheimpflug Corneal Analyzer (A) examiner view, (B) patient view, (C) summary report.

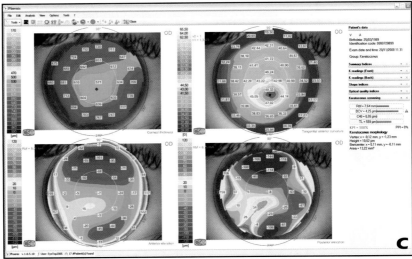

CSO Sirius

An Italian company (CSO, Costruzione Strumenti Oftalmici, Florence, Italy) has also introduced a Scheimpflug camera instrument that has also been shown to be accurate and repeatable (Fig. 32-9).

Methods to Adjust/Calculate the Target Intraocular Lens Power

Those Needing Clinical History

Aramberri Double-K Method[32]

Use K_{PRE} to calculate ELP and K_{PO} to calculate IOL power.

One of the most important developments to improve the prediction of corneal power in eyes that have had refractive surgery was proposed in 2001 and is termed the "Double-K" method by Jaime Aramberri of San Sebastian, Spain. His proposal makes eminent sense. The modern theoretic formulas (except the Haigis) use the input of corneal power for two purposes: the first is to predict the ultimate position of the IOL (ACD or ELP) and the second (along with AL, target refraction, and ELP) is to calculate the power of the IOL. The formulations and algorithms used to predict the ELP are based on the anatomy of the anterior segment, which is not changed by corneal refractive surgery (only the center is flattened and thinned). Therefore, if the postoperative refractive surgery K reading (which is flatter) is used to calculate the ELP, it will produce an erroneous ELP value. Because the anatomy has not changed, Aramberri recommends the use of the preoperative K reading to calculate the ELP. The IOL power is then calculated using the postoperative K reading, thus the use of two K readings ("Double-K"). His analysis of a small series of eyes proved the benefit of this idea.

Feiz-Mannis Formula[33]

$$P = P_E - RC_S/0.7$$

In this method you calculate the emmetropic IOL power using the preoperative K reading and adjust that value (P_E) using the surgically induced refractive change.

Feiz-Mannis Method[34]

This method utilizes the change in refractive error to offset the calculated target IOL power. There is one formula for myopic eyes and another for hyperopic:

Myopic Eye $\quad P = P_{TARG} - 0.595 * RC_C + 0.231$

Hyperopic Eye $\quad P = P_{TARG} - 0.862 * RC_C + 0.751$

Latkany Methods[35]

Myopic Eye $\quad P = P_{TARG\ FlatK} - 0.47 * R_{PRE} + 0.85$

Hyperopic Eye $\quad P = P_{TARG\ FlatK} + 0.27 * R_{PRE} + 1.53$

This method requires knowledge of the pre-LASIK refractive error and the calculation of the target IOL power using the flattest postoperative K rather than the usual average K.

Masket Method[36]

$$P = P_{TARG} - 0.323 * RC_C + 0.138$$

[SRK/T: myopes; Hoffer Q: hyperopes]

Table 32-2.

EXAMPLE CALCULATIONS USING THE MASKET FORMULA

Myopic Eye	Hyperopic Eye
SRK/T calculates 16.0 D IOL	Hoffer Q calculates 22.0 D
Change in Rx = -6.0 D	Change in Rx = +3.0 D
-0.323 * (-6) + 0.138 = +2.076	-0.323 * (+3) + 0.138 = -0.82
P = 16.0 + 2.0 = 18.0 D	P = 22.0 − 1.0 = 21.0 D

This method is a play on the Latkany method, which adjusts the power of the IOL calculated using the postoperative measured data and the knowledge of the surgically induced refractive change. He recommends using the SRK/T formula for myopic ALs and the Hoffer Q for hyperopic ALs. Example calculations are shown in Table 32-2.

In a series of 28 post-LASIK eyes, he reported 43% of the eyes obtaining a postoperative refractive error of plano, 95% within ±0.50 D of prediction, and a total error range from -0.75 D to +0.50 D.

Wake Forest Method[37]

Use R_{PRE} as the RX_{TARG} using measured AL and K_{PRE}

In 2005, Gagnon et al (from Wake Forest University) published an alternative calculation method that has been discussed by others over the past 20 years. This method simply uses the patient's preoperative refraction before LASIK as the target or "desired" PO refraction in the calculation and the measured AL and K readings without modification.

THOSE NOT NEEDING CLINICAL HISTORY

Aramberri Double-K Method[32]

Use 43.5 or 44.00 to calc ELP & K_{PO} to calc IOL power.

The use of a standard normal K reading in the Double-K method is a great improvement over using the calculated very flat K reading.

Ianchulev Intraoperative Aphakic Refraction Method[38]

$$P = 2.02 * AR + (A - 118.4)$$

In 2003, Ianchulev et al proposed calculating IOL power by performing an aphakic refraction on the operating room table using a hand-held automated refractor immediately after the cataract has been removed and the AC is inflated to normal status. The resultant refraction is modified by the formula.

His early results are quite promising (see Chapter 44). This method would completely eliminate the need for axial length, corneal power measurements, and the problems with LASIK and silicone oil-filled eyes. However, it would require a large IOL inventory in the operating room.

Mackool Secondary Implant Method[39]

$$P = 1.75 * AR + (A - 118.84)$$

This method is similar to the previous method, except the patient is removed from the operating room without an IOL implanted, then refracted in a refraction lane and taken back to the operating room for secondary lens implantation. It is my impression that this method would not be popular with most surgeons.

Formula Legends

A = the IOL A constant for planned IOL style
AL = axial length
AR = aphakic refractive error (SE)
B = base curve, PCL = power of CL, NoCL = bare refraction
CL = contact lens
K = predicted PO corneal power
K_{PO} = the average PO corneal power by manual keratometry (in diopters D)
K_{POFLAT} = flattest measured PO manual keratometry
K_{POr} = the average PO corneal power by IOLMaster (in radius r [mm])
K_{PRE} = refractive surgery preoperative corneal power (K readings)
P = IOL Power
P_{EMM} = the IOL power calculated for emmetropia
P_{FlatK} = IOL power calculated for Rx_{TARG} using the PO flattest K
P_{TARG} = the target IOL power to produce the PO desired refractive error
R = refractive error: $_{PRE}$ = preoperative, $_{PO}$ = postoperative
RC_C = surgical change in refractive error (SE) vertexed to corneal plane
RCF = Rosa Correction Factor based on axial length
RC_S = surgical change in refractive error (SE) at spectacle plane
R_{PO} = refractive surgery PO refractive error (spherical equivalent)
R_{PRE} = refractive surgery preoperative refractive error (spherical equivalent)
Rx_{TARG} = planned postoperative target refractive error
TK = average PO topography central Sim-K or EffRP
TK_{CTR} = exact singular PO topography central K

References

1. Koch DD, Liu JF, Hyde LL, Rock RL, Emery JM. Refractive complications of cataract surgery after radial keratotomy. *Am J Ophthalmol.* 1989;108(6):676–682.
2. Hoffer KJ. The Hoffer Q formula: A comparison of theoretic and regression formulas. *J Cataract Refract Surg.* 1993;19(6):700–712. Errata: 1994;20(6):677 and 2007;33(1):2–3.
3. Holladay JT, Prager TC, Chandler TY, et al. A three-part system for refining intraocular lens power calculations. *J Cataract Refract Surg.* 1988;14(1):17–24.
4. Retzlaff J, Sanders DR, Kraff MC. Development of the SRK/T intraocular lens implant power calculation formula. *J Cataract Refract Surg.* 1990;16(3):333–340. Erratum: 1990;16(4):528.
5. Haigis W. The Haigis formula. In: HJ Shammas, ed. *Intraocular Lens Power Calculations.* Thorofare, NJ: SLACK Incorporated; 2003:41–57.
6. Holladay JT. IOL calculations following radial keratotomy surgery. *Refract Corneal Surg.* 1989;5:36A.

7. Hoffer KJ. Intraocular lens power calculation for eyes after refractive keratotomy. *J Refract Surg.* 1995;11(6):490–493.
8. Hoffer KJ. Calculating intraocular lens power after refractive surgery. *Arch Ophthalmol.* 2002;120(4):500–501.
9. Ridley F. Development in contact lens theory. *Trans Ophthalmol Soc UK.* 1948;68:385–401.
10. Soper JW, Goffman J. Contact lens fitting by retinoscopy. In: Soper JW, ed. *Contact Lenses.* New York, NY: Stratton Intercontinental Medical Book Corp.; 1974;99.
11. Odenthal MTP, Eggink CA, Melles G, et al. Clinical and theoretical results of intraocular lens power calculation for cataract after photorefractive keratectomy for myopia. *Arch Ophthalmol.* 2002;120(4):431–438.
12. Hamed AM, Wang L, Misra M, Koch D. A comparative analysis of five methods of determining corneal refractive power in eyes that have undergone myopic laser in situ keratomileusis. *Ophthalmology.* 2002;109(4):651–658.
13. Speicher L. Intra-ocular lens calculation status after corneal refractive surgery. *Curr Opin Ophthalmol.* 2001;12(1):17–29.
14. Seitz B, Langenbucher A, Nguyen NX, Kus MM, Kuchle M. Underestimation of intraocular lens power for cataract surgery after myopic photorefractive keratectomy. *Ophthalmology.* 1999;106(4):693–702.
15. Seitz B, Langenbucher A. Intraocular lens power calculation in eyes after corneal refractive surgery. *J Refract Surg.* 2000;16(3):349–361.
16. Jarade EF, Abi Nader FC, Tabbara KF. Intraocular lens power calculation following LASIK: Determination of the new effective index of refraction. *J Refract Surg.* 2006;22(1):75–80.
17. Ronje. LASIK IOL calculation. *Eyenet Magazine.* 2004;20:23–24.
18. Savini G, Barboni P, Zanini M. Intraocular lens power calculation after myopic refractive surgery: Theoretical comparison of different methods. *Ophthalmology.* 2006;113(8):1271–1282.
19. Camellin M, Calossi A. A new formula for intraocular lens power calculation after refractive corneal surgery. *J Refract Surg.* 2006;22(2):187–199.
20. Jarade EF, Tabbara KF. New formula for calculating intraocular lens power after laser in situ keratomileusis. *J Cataract Refract Surg.* 2004;30(8):1711–1715.
21. Smith RJ, Chan WK, Maloney RK. The prediction of surgically induced refractive change from corneal topography. *Am J Ophthalmol.* 1998;125(1):44–53.
22. Koch D, Wang I. Calculating IOL power in eyes that have had refractive surgery. *J Cataract Refract Surg.* 2003;29(11):2039–2042.
23. Savini G, Barboni P, Zanini M. Correlation between attempted correction and keratometric refractive index of the cornea after myopic excimer laser surgery. *J Refract Surg.* 2007;23(5):461–466.
24. Shammas HJ, Shammas MC, Garabet A, Kim JH, Shammas A, LaBree L. Correcting the corneal power measurements for intraocular lens power calculations after myopic laser in situ keratomileusis. *Am J Ophthalmol.* 2003;136(3):426–432.
25. Ferrara G, Cennamo G, Marotta G, Loffredo E. New formula to calculate corneal power after refractive surgery. *J Refract Surg.* 2004;20(5):465–471.
26. Rosa N, Capasso L, Lanza M, Iaccarino G, Romano A. Reliability of a new correcting factor in calculating intraocular lens power after refractive corneal surgery. *J Cataract Refract Surg.* 2005;31:1020-1024.
27. Haigis W. IOL calculation after refractive surgery for myopia: The Haigis-L formula. *J Cataract Refract Surg.* 2008;34(10):1658–1663.
28. Savini G, Barboni P, Carbonelli M, Hoffer KJ. Agreement between Pentacam and videokeratography in corneal power assessment. *J Refract Surg.* 2009;25:534-538.
29. Savini G, Barboni P, Carbonelli M, Hoffer KJ. Accuracy of Scheimpflug corneal power measurements for intraocular lens power calculation. *J Cataract Refract Surg.* 2009;35(7):1193-1197.

30. Tang Q, Hoffer KJ, Olsen MD, Miller KM. Accuracy of Scheimpflug Holladay equivalent keratometry readings after corneal refractive surgery. *J Cataract Refract Surg.* 2009;35(7):1198-1203.
31. Borasio E, Stevens J, Smith GT. Estimation of true corneal power after keratorefractive surgery in eyes requiring cataract surgery: BESSt formula. *J Cataract Refract Surg.* 2006;32(12):2004–2014.
32. Aramberri J. Intraocular lens power calculation after corneal refractive surgery: double-K method. *J Cataract Refract Surg.* 2003;29(11):2063–2068.
33. Feiz V, Mannis MJ, Garcia-Ferrer F. Intraocular lens power calculation after laser in situ keratomileusis for myopia and hyperopia: A standardized approach. *Cornea.* 2001;20(8):792–797.
34. Feiz V, Moshirfar M, Mannis MJ, et al. Nomogram-based intraocular lens power adjustment after myopic photorefractive keratectomy and LASIK: A new approach. *Ophthalmology.* 2005;112(8):1381–1387.
35. Latkany RA, Chokshi AR, Speaker MG, Abramson J, Soloway BD, Yu G. Intraocular lens calculations after refractive surgery. *J Cataract Refract Surg.* 2005;31(3):562–570.
36. Masket S, Masket SE. Simple regression formula for intraocular lens power adjustment in eyes requiring cataract surgery after excimer laser photoablation. *J Cataract Refract Surg.* 2006;32(3):430–434.
37. Walter KA, Gagnon MR, Hoopes PC Jr., Dickenson PJ. Accurate intraocular lens power calculation after myopic laser in situ keratomileusis, bypassing corneal power. *J Cataract Refract Surg.* 2006;32(3):425–429.
38. Ianchulev T, Salz J, Hoffer K, et al. Intraoperative optical intraocular lens power estimation without axial length measurements. *J Cataract Refract Surg.* 2005;31(8):1530–1536.
39. Mackool RJ, Ko W, Mackool R. Intraocular lens power calculation after laser in situ keratomileusis: The aphakic refraction technique. *J Cataract Refract Surg.* 2006;32(3):435–437.

Special Circumstances: Haigis-L IOL Formula

Wolfgang Haigis, MS, PhD

The number of patients presenting with cataract after refractive corneal surgery has been continuously increasing over the years. These patients still present a challenge to IOL calculation, although the special problems associated with these eyes are well understood today.

Problems for IOL Calculation

Essentially there are 3 sources of errors in IOL calculation after refractive surgery. First, there is the radius measurement error stemming from the fact that K readings are not taken at the optical axis, but a little peripherally. This error is relevant in cases of preceding laser surgery for myopia, not for hyperopia. Second, the keratometer index error is due to the fact that the corneal curvature ratio is deliberately altered by refractive surgery, thus leading to meaningless K values. Third, some IOL power formulas making use of K values to predict the effective lens position derive a wrong value since the current K does not represent the eye's geometry anymore, as it does in its untouched state. This IOL formula error causes a hyperopic refractive shift in patients after laser surgery for myopia.

Approaches in the Literature

A variety of approaches to handle eyes after previous refractive surgery can be found in the literature (see Chapter 32). Formulas to estimate the effective corneal power, as well as formulas to fudge the calculated IOL power, are available. Often historical data, additional measurements, and/or special measurement parameters are required. There is no magic general formula valid for all cases; usually solutions are described for specific measurement instruments and/or specific IOL power formulas. IOL calculation methods for eyes after refractive surgery differ most in whether they require historical patient data or whether they rely only on current measurements. Obviously, no-history methods are clinically the most useful ones. Among them are the No-History method of Shammas, the Pentacam-based BESSt formula of Borasio, and the Haigis-L formula[1] for the IOLMaster (see Chapter 30). A spreadsheet (Hoffer/Savini Tool) programmed with virtually all algorithms hitherto published is available for download from www.EyeLab.com at no cost. Also, on the ASCRS website (www.ascrs.org) an online calculator is implemented offering the free use of a limited variety of published calculation schemes, but there are limitations as to the formulas used.

The Haigis-L Formula

The Haigis-L formula consists of the regular Haigis formula[2] and a special correction for the corneal radius (r_{meas}) measured by the Zeiss IOLMaster:

$$r_{corr} = \frac{331.5}{-5.1625 * r_{meas} + 82.2603 - 0.35}$$

where r = IOLMaster radius of curvature of the cornea, corr = corrected and meas = measured.

It is important to remember this correction is specific only for the Zeiss IOLMaster.

Clinical Results With the Haigis-L Formula

At present, we are studying the clinical results with the Haigis-L formula specific for the IOLMaster; for 222 eyes after IOL implantation with previous myopic and 56 with previous hyperopic laser vision correction. The previously myopic eyes received 35 different IOL types by 64 different surgeons from all over the world; former hyperopic eyes were implanted with 13 different IOL types by 15 different surgeons. All patients had biometry and keratometry using the Zeiss IOLMaster. IOL calculation was performed from current measurements using the Haigis-L formula (which is included in the IOLMaster software).

The mean arithmetic prediction errors (ME) were -0.08 ±0.71 D for myopic and -0.06 ±0.77 D for hyperopic eyes. The respective median absolute errors (MAE) were 0.37 D and 0.40 D. Of the myopic eyes, 98.6% were correctly predicted within ±2 D, 82.9% within ±1 D, and 59.9% within ±0.5 D. The respective percentages for eyes after surgery for hyperopia were 96.4%, 82.1%, and 58.9%. These results compare well with normal eyes, although the error ranges in predicted refraction are a little higher in eyes after refractive surgery.

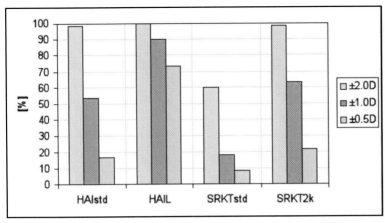

Figure 33-1. Percentages of correct refraction predictions obtained with the standard Haigis (HAIstd) and Haigis-L formulas (HAIL), as well as the standard SRK/T (SRKTstd) and the Double-K corrected (SRKT2k) SRK/T formulas for 60 eyes with an Alcon SN60WF with previous laser surgery for myopia (unpublished data).

SRK/T[3] is the IOL power formula most affected by the formula error. The formula error can be lessened by applying the Aramberri[4] Double-K correction. The Hoffer Q formula is least effected, but the Haigis formula, on the other hand, does not suffer from this error because it does not use the K reading as a predictor for the effective lens position (ELP).

Fig. 33-1 shows the performances of the Haigis and SRK/T formulas expressed in correct refraction predictions in their standard versions as well as in their adapted versions for preceding refractive surgery (Haigis-L and SRK/T with Double-K correction) for 60 eyes with an Alcon SN60WF. The Haigis-L formula, as can be seen, compares very well with the other calculation approaches.

References

1. Haigis W. IOL calculation after refractive surgery for myopia: The Haigis-L formula. *J Cataract Refract Surg.* 2008;34(10):1658-1663.
2. Haigis W, Lege B, Miller N, Schneider B. Comparison of immersion ultrasound biometry and partial coherence interferometry for intraocular lens calculation according to Haigis. *Graefe's Arch Clin Exp Ophthalmol.* 2000;238:765-773.
3. Retzlaff J, Sanders DR, Kraff MC. Development of the SRK/T intraocular lens implant power calculation formula. *J Cataract Refract Surg.* 1990;16(3):333-340.
4. Aramberri J. Intraocular lens power calculation after corneal refractive surgery: Double K method. *J Cataract Refract Surg.* 2003;29(11): 2063-2068.

Special Circumstances: Double-K Method

Jaime Aramberri, MD

All theoretical IOL power calculation formulas perform two consecutive calculations: first the position of the IOL within the eye (ELP: effective lens position in thin lens terminology[1]) is estimated from different independent variables; and then the power of the implant is calculated using optical vergence or ray tracing formulation.

Since the introduction of 3rd generation formulas in 1988,[2] corneal curvature has been used as a powerful ELP predicting variable in most of them; the steeper the cornea the higher the ELP, and therefore the higher the IOL power. However this anatomical correlation fails in abnormally steep or flat corneas, the latter due to corneal refractive surgery eyes. The Double-K Method avoids the ELP prediction error in these eyes.

It must also be remembered that there is another source of error after corneal refractive surgery which is the K measurement error. Topographers and keratometers overestimate K value after myopic surgery and underestimate K value after hyperopic surgery due to the altered corneal anterior/posterior ratio. This error and its correction will be explained elsewhere.

Corneal Refractive Surgery Anatomical Changes

Laser corneal refractive surgery (LASIK/PRK) flattens the anterior corneal surface but does not change the posterior surface according to Scheimpflug measurements or steepen it according to Orbscan measurements. The latter has been reported to be an artefact due to error in boundaries recognition or magnification.[3] Anterior chamber depth decreases a small and nonsignificant amount.[4]

Table 34-1.

ELP Predicting Variables Used by Theoretic Formulas

Formula	K	AL	ACD	LT	CD	Rx	Age
Binkhorst 2	No	Yes	No	No	No	No	No
SRK/T	Yes	Yes	No	No	No	No	No
Hoffer Q	Yes	Yes	No	No	No	No	No
Holladay 1	Yes	Yes	No	No	No	No	No
Holladay 2	Yes	Yes	Yes	Yes	Yes	Yes	Yes
Haigis	No	Yes	Yes	No	No	No	No
Olsen	Yes	Yes	Yes	Yes	No	Yes	No

K = corneal power, AL = axial length, ACD = preoperative anterior chamber depth (anterior cornea to anterior lens), LT = lens thickness, Rx = Refraction, and CD = corneal diameter (horizontal white-to-white distance.)

ELP Prediction

Third generation formulas use 2 independent variables to predict ELP (Table 34-1). Hoffer Q,[2] Holladay 1,[5] and SRK/T[6] use AL and K. Haigis[7] uses AL and preoperative ACD (anterior corneal vertex to anterior lens). Fourth generation formulas use more than 2 variables for the same task. The Holladay 2[8] uses 7: AL, K, ACD, lens thickness (LT), corneal diameter (CD), refraction (Rx), and age. Olsen[9] (2006) uses 5 variables: AL, K, ACD, LT, and Rx. The main difference among all formulas is the prediction of ELP. In fact, if ELP is fixed to test the optical formula performance it can be seen that Hoffer Q, Holladay 1, and SRK/T predict IOL power within 0.5 D from each other. Haigis always calculates 0.50 to 1.00 D higher power (Fig. 34-1). This difference is systematic so it can have an effect on the a_0, a_1, and a_2 values used for the calculations. Knowledge of ELP prediction performance of each formula helps in understanding the potential sources of error.

SRK/T limits ELP prediction neither superior nor inferiorly (Fig. 34-2). High K and AL values will produce illogically high ELP values, ie, ELP for K = 45 and AL = 30 is 7.65 mm which of course is senseless; no IOL sits that deeply. This will increase IOL power in a magnitude dependant on the power of the IOL itself. A curious phenomenon is a sudden decrease of ELP for K values over 46. The higher the K is (over 46 D) the lower the AL needed to produce this prediction change. The consequence is that SRK/T predicts ELP more accurately in long eyes when the K is very high (>47) than when the K is in the mid-high range (45 to 47). For very low K values, SRK/T predicts the lowest ELP value of all.

Haigis predicts ELP linearly as a function of AL for a fixed ACD despite variable K values (which is not a predicting variable as has been stated). This function is not limited either superior or inferiorly (Fig. 34-3).

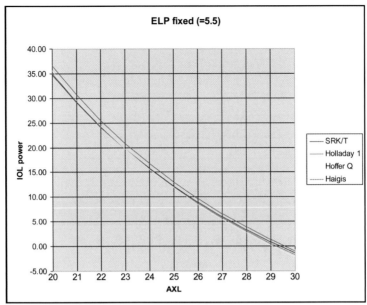

Figure 34-1. IOL power prediction of 3rd generation formulas for a fixed ELP value of 5.50 mm. Difference among Hoffer Q, Holladay 1, and SRK/T is lower than 0.5 D. Haigis formula calculates a higher value (0.50 to 1.00 D) throughout the range of AL.

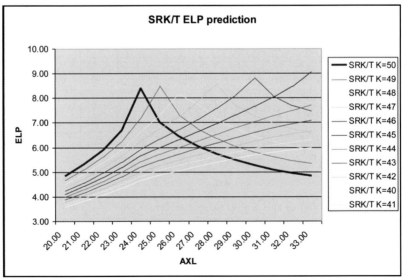

Figure 34-2. SRK/T ELP prediction. When K >46, an abrupt fall of ELP occurs. High K and AL values lead to an overestimation of ELP. ELP values above 7 mm are seldom found in pseudophakic eyes.

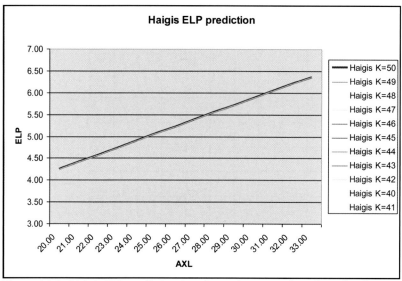

Figure 34-3. Haigis ELP prediction is a linear function of AL for a fixed ACD. The K is not a predicting variable. There are no upper or lower limits in this function.

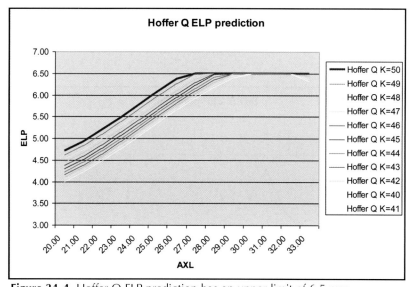

Figure 34-4. Hoffer Q ELP prediction has an upper limit of 6.5 mm.

Hoffer Q (Fig. 34-4) and Holladay 1 (Fig. 34-5) algorithms limit superior ELP prediction to avoid ELP overestimation in long eyes and/or steep corneas. The Holladay 1 limits ELP over 26 mm of AL in a value that depends on K. The Hoffer Q sets an absolute ELP limit of 6.50 mm. With very low K values, SRK/T predicts the shortest ELP value and Hoffer Q the highest (Fig. 34-6) with significant differences.

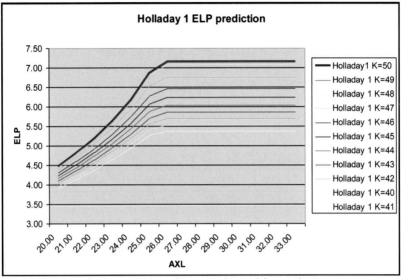

Figure 34-5. Holladay 1 ELP prediction is limited beyond AX = 26 mm in a value that depends on K.

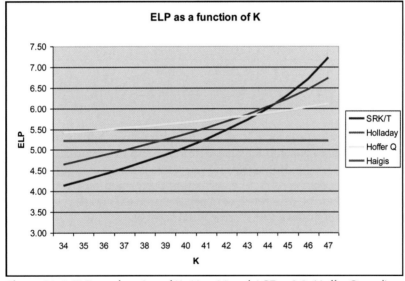

Figure 34-6. ELP as a function of K. AL = 26 and ACD = 3.3. Hoffer Q predicts the highest values for very flat corneas. This is why Single-K Hoffer Q induces less hyperopia than Single-K Holladay 1 and SRK/T after myopic corneal refractive surgery. Haigis is not affected by K as it is not a predicting variable

ELP Prediction After Corneal Refractive Surgery

As the relationship between corneal curvature and pseudophakic ACD has changed after corneal refractive surgery, it is obvious that using the post-surgical K value to predict that variable will lead to an error in any algorithm that uses K in ELP prediction. After myopic surgery, the new flat K value will induce an underestimation of ELP and

Table 34-2.

RESULTS OF ELP ERROR OF HOFFER Q, HOLLADAY 1, AND SRK/T*

*Assuming that ELP estimation of the unchanged cornea (First row: K = 44, Refractive correction [Rx] = 0) is correct. The highest error is induced by the SRK/T algorithm (1.72), whereas the Hoffer Q introduces the least error (0.44); with Holladay 1 in between (1.27).

AL	K_{pre}	Rx		Hoffer Q		Holladay 1		SRK/T	
mm	D	D		ELP mm	ELP error	ELP mm	ELP error	ELP mm	ELP error
26	44	0		5.88	0.00	6.04	0.00	5.99	0.00
26	43	1		5.82	0.06	5.85	0.18	5.72	0.28
26	42	2		5.76	0.12	5.69	0.35	5.47	0.52
26	41	3		5.71	0.18	5.53	0.51	5.26	0.73
26	40	4		5.65	0.23	5.38	0.65	5.06	0.93
26	39	5		5.61	0.28	5.25	0.79	4.88	1.11
26	38	6		5.56	0.32	5.12	0.92	4.71	1.28
26	37	7		5.52	0.36	5.00	1.04	4.55	1.44
26	36	8		5.48	0.40	4.88	1.16	4.41	1.58
26	35	9		5.44	0.44	4.77	1.27	4.27	1.72

therefore an underestimation of IOL power resulting in a hyperopic refraction. After hyperopic surgery, the new steep K value will induce an overestimation of ELP, an overestimation of IOL power yielding a myopic refractive error.

This error depends on the refractive correction performed on the cornea, because the higher this is the higher the difference between the original and the post surgical K value is and consequently the induced ELP error will increase.

Another factor is which formula is being used because ELP prediction algorithms are different, as stated above. Table 34-2 shows that SRK/T induces the biggest error and Hoffer Q the lowest (with a very significant difference). The reason is that the slopes of these functions are very different; SRK/T decreases ELP ~0.2 mm per D of K, Holladay 1 ~0.15 mm per D of K, and the Hoffer Q only ~0.05 mm per D of K. This means, in the case of a 7 D post-LASIK calculation with AL = 26 mm, SRK/T will underestimate ELP by 1.44 mm which will translate into 1.4 D of hyperopia, the Holladay 1 will underestimate ELP by 1.04 mm resulting in 1 D of hyperopia, and the Hoffer Q will underestimate ELP by 0.36 mm producing only 0.30 D of hyperopia (approximate numbers).

The magnitude of IOL power and spectacle plane refractive error produced by any ELP error will depend mainly on AL (Table 34-3) and to a lesser extent on K value.

Table 34-3.

IOL Error and Spectacle Plane Refractive Error Produced by a 0.50 mm Error in ELP Prediction

The refractive translation of this error is clearly dependant on AL. Calculations were performed by paraxial ray tracing.

AL (mm)	IOL Error (D)	Rx Error* (D)
21	1.76	1.23
23	1.10	0.80
27	0.70	0.50
30	0.40	0.20

*Spectacle plane refractive error. Calculations performed by paraxial ray tracing.

ELP Prediction Error Correction

The easiest way to correct this problem is not using the K as an ELP predicting variable. The only formula programmed in this way is the Haigis formula, which estimates ELP using only AL and ACD.

If a formula that uses corneal power (K or r) for ELP prediction is used, the formula must be programmed in such a way that the K previous to corneal refractive surgery (K_{pre}) is input into the ELP predicting algorithm and the K after corneal refractive surgery (K_{post}) is input into the optical power formula. This K_{post} is not the value measured by the keratometer or topographer but a corrected K value that can be calculated using different methods as is explained elsewhere in this book (see Chapter 30). The use of two different K values in IOL power calculation has been called the Double-K Method.[10]

In the original published paper,[10] the Double-K SRK/T has been recommended to be the more accurate in long eyes. But since then I have discovered (nonpublished data) that this formula overestimates ELP, especially if K_{pre} is between 44 and 46 D—inducing some myopia. This behavior was not so obvious with nonoperated eyes, as the IOL power prediction difference is low with low-powered IOLs (needed in myopes). Better 3rd generation options are Double-K Hoffer Q and Double-K Holladay 1.

How to Obtain Double-K Formula Calculations

All formulas (except Holladay 2) have been published in the scientific peer-reviewed literature. It is advisable to get the articles and program them in a spreadsheet in such a way that the ELP-predicting algorithm uses K_{pre} as the independent variable and the optical vergence formula uses K_{post} as the independent variable. It is important to be aware of the crucial errata in the Hoffer Q and SRK/T publications. Conversion tables have been published[11] to translate Single-K calculations to Double-K ones.

Some ultrasonic biometers have included Double-K formulas in their software, such as Axis II (Quantel) and Sonomed A-Scans. Commercial IOL power calculation software include Double-K formulas: Hoffer Programs® (www.eyelab.com) allows using this method with all 3rd generation formulas. The Holladay IOL Consultant® program (www.docholladay.com) uses this method only with the Holladay 2 formula (unpublished data).

References

1. Holladay JT. Standardizing constants for ultrasonic biometry, keratometry, and intraocular lens calculation. *J Cataract Refract Surg.* 1997;23(9):1356–1370.
2. Holladay JT, Prager TC, Chandler TY, et al. A three-part system for refining intraocular lens power calculations. *J Cataract Refract Surg.* 1988;14(1):17–24.
3. Nawa Y, Masuda K, Ueda T, Hara Y, Uozato H. Evaluation of apparent ectasia of the posterior surface of the cornea after keratorefractive surgery. *J Cataract Refract Surg.* 2005;31(3):571–573.
4. Hashemi H, Mehravaran S. Corneal changes after laser refractive surgery for myopia: Comparison of Orbscan II and Pentacam findings. *J Cataract Refract Surg.* 2007;33(5):841–847.
5. Hoffer KJ. The Hoffer Q formula: A comparison of theoretic and regression formulas [published correction appears in: *J Cataract Refract Surg.* 1994;20(6):677 and *J Cataract Refract Surg.* 2007;33(1):2–3]. *J Cataract Refract Surg.* 1993;19(6):700–712.
6. Retzlaff J, Sanders DR, Kraff MC. Development of the SRK/T intraocular lens implant power calculation formula [published correction appears in: *J Cataract Refract Surg.* 1990;16(4):528]. *J Cataract Refract Surg.* 1990;16(3):333–340.
7. Haigis W. IOL calculation according to Haigis. Available online: htpp://www.augenklinik.uni-wuerzburg.de/uslab/ioltxt/haie.htm. Last revision: December 7, 1998.
8. Hoffer KJ. Clinical results using the Holladay 2 intraocular lens power formula. *J Cataract Refract Surg.* 2000;26(8):1233–1237.
9. Olsen T. Prediction of the effective postoperative (intraocular lens) anterior chamber depth. *J Cataract Refract Surg.* 2006;32(3):419–424.
10. Aramberri J. Intraocular lens power calculation after corneal refractive surgery: Double-K method. *J Cataract Refract. Surg.* 2003;29(11):2063–2068.
11. Koch DD, Wang L. Calculating IOL power in eyes that have had refractive surgery. *J Cataract Refract Surg.* 2003;29(11):2039–2042.

35

Special Circumstances: Influence of Spherical Aberration on IOL Power

Sverker Norrby, PhD

As the name suggests, spherical aberration is a property of spherical lenses. For a postive spherical lens, peripheral rays are refracted more than rays close to the axis (positive spherical aberration). How much depends on the curvature of the refracting surfaces and the shape of the lens.

In Table 35-1, dimensions (mm) are shown for different shapes of 20 D IOLs made of a material with a refractive index (RI) = 1.460, when immersed in aqueous (RI = 1.336). Also included are the parameters for the Gullstrand cornea. Note that radii convex towards the object are positive, and those concave are negative. Plano surfaces are given a very large radius.

The values for the equi-convex lens are input into the calculation spreadsheet (Figs. 35-1 and 2), which can be downloaded from AAO at http://one.aao.org/lms/courses/IOLPowerFormulas/images/LO14_GC01.xls *and* http://one.aao.org/lms/courses/IOLPowerFormulas/images/LO14_GC02.xls. The marginal ray is traced at the edge of the aperture. The focusing ray is traced at $1/\sqrt{2}$ of the aperture. This radius divides the pupil into an inner circle and an outer annulus of equal area. The focusing ray can thus be considered an average ray and its focal point can be considered as "best focus" in that sense. The tracing of these two rays is exact. The paraxial ray is traced at an incoming height of 1 mm for visibility. Paraxial ray tracing is an approximation and the focal point is independent of the height chosen. This can be demonstrated by giving it other values in the spreadsheet.

The distances are termed as follows:
- From back vertex of lens to best-focus: back focal length (BFL)

Table 35-1.

DIMENSIONS (MM) OF DIFFERENT SHAPES OF 20 D IOLS

Lens shape	Equi-convex	Convex-plano	Plano-convex	Meniscus	Cornea
Front radius	12.336	6.200	1000000	3.876	7.70
Back radius	-12.336	-1000000	-6.200	10.000	6.80
Thickness	1.50	1.50	1.50	1.50	0.50
Diameter	6.00	6.00	6.00	6.00	12.00
Object RI	1.336	1.336	1.336	1.336	1
Image RI	1.336	1.336	1.336	1.336	1.336
Lens RI	1.460	1.460	1.460	1.460	1.376

where RI = refractive index.

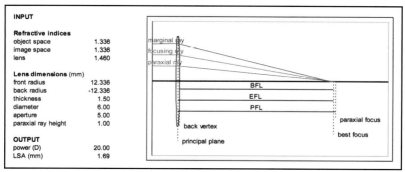

Figure 35-1. Graphic Calculator Ray Tracing Spreadsheet: Paraxial thick lens ray tracing of single lenses. The spreadsheet is interactive and will accept any input, though the graph may not remain within the limits for excessive numbers.

- From principal plane to best-focus: effective focal length (EFL)
- From principal plane to paraxial focus: paraxial focal length (PFL)

IOLs should, in accordance with the international standard (ISO 11979-2), be labeled with their paraxial power, which is 1336/PFL with PFL given in mm. The distance between marginal focus and paraxial focus is the longitudinal spherical aberration (LSA), which is output in the spreadsheet.

Input the values for the other lens shapes and watch the consequences. Also try the IOLs in air (object and image RI = 1) and see how much higher the power is in air. Also try the cornea. Note in particular how the principal plane shifts with shape. For equi-convex lenses it is slightly posterior to the middle, for convex-plano lenses it is a little posterior to the anterior vertex. For plano-convex lenses it is exactly at the posterior vertex, and for meniscus lenses (including the cornea) it is slightly anterior to the anterior vertex.

Also see how much influence shape has on LSA and how this influence is different in air and in aqueous. The influences of shape are different with the lens in the converging light behind the cornea. Therefore, no inference can be made on how an IOL will perform in the eye from its performance in isolation in neither air nor aqueous. In

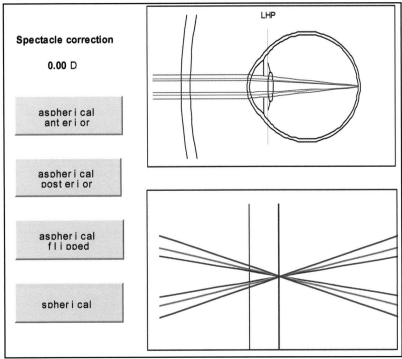

Figure 35-2. Graphic Calculator Ray tracing of a normal eye with a 5 mm pupil, with marginal (blue), focusing (green), and paraxial (red) rays drawn. Dimensions are chosen to exactly focus with a 20 D fully correcting aspherical IOL.
Top view: Entire eye with spectacle. The lens haptic plane (LHP) is shown.
Bottom view: Focusing of rays at the receptor plane (brown line). The thin black line depicts the inner limiting membrane 0.25 mm anterior to the receptor plane of the retina. Activation of the spreadsheet allows 4 cases to be illustrated. Try the buttons on the spreadsheet and see how spectacle power and rays at the retina change when the optic is reversed, the IOL is flipped around LHP, or when the lens is spherical.

the pseudophakic eye it is the combined spherical aberration of the cornea and the IOL that matters.

Popular IOL power formulas use thin lens theory, in which power is associated with the principal plane of lenses. However, the power in the principal plane of the cornea is about 1.5 D less in its principal plane than given by most keratometers.[2]

The principal plane of the IOL shifts anteriorly in the converging light behind the cornea.[3] The AL measured by A-scan ultrasound (and also by the IOLMaster) is shorter than the distance from the corneal principal plane to the image plane on the photo receptors, which is the distance that should be used in an optical calculation. With normal (positive) spherical aberration, the average pseudophakic eye is approximately 0.5 D stronger than without spherical aberration for a pupil size of 4 mm. All these differences are systematic and are absorbed in the IOL constants that apply to the power formulas. Because both keratometers and AL measuring instruments can differ systematically from each other, IOL constants should be "personalized" to give, on the average, zero error in refractive outcome in a given setting.

Aspheric IOLs that correct all or part of the spherical aberration of the cornea require adjusted formula constants to give the desired refractive outcome. Therefore an aspheric

IOL with otherwise the same design as its spherical counterpart, will have formula constants that will result in higher powers being indicated by the calculation.

Activation of the spreadsheet allows 4 cases to be illustrated. Try the buttons on the spreadsheet and see how spectacle power and rays at the retina change when the optic is reversed, the IOL is flipped around LHP, or when the lens is spherical.

References

1. ISO 11979-2. Ophthalmic implants—Intraocular lenses—Part 2: Optical properties and test methods. Geneva, Switzerland: International Organization for Standardization; 2000.
2. Norrby S. Letter: Pentacam keratometry and IOL power calculation. *J Cataract Refract Surg.* 2008;34(1):3;reply 4.
3. Holladay JT, Maverick KJ. Relationship of the actual thick intraocular lens optic to the thin lens equivalent. *Am J Ophthalmol.* 1998;126(3):339–347.

Special Circumstances: Multifocals and Toric IOLs

John Moran, MD, PhD

Calculating the optimal power of toric and multifocal IOLs can be viewed as an extension and refinement of the methods used to calculate the power of spherical IOLs. The same principles of minimizing systematic and random errors by using consistent biometric methods and surgical techniques along with optimized lens constants and a modern power formula (Haigis, Hoffer Q, Holladay, SRK/T) apply. Additionally, the desired postoperative distant and near refractive targets, as well as the surgically induced astigmatism, require careful consideration to achieve optimal results.

Multifocal IOLs

The IOL model and power should be selected to meet the patient's most important distance and near visual tasks. For example, a cellist reading sheet music and a seamstress require different near points.

IOLs are available with different strength adds. Two recently introduced IOLs, the Alcon Acrysof Restor (Alcon USA, Fort Worth, TX) (Fig. 36-1) and the AMO Rezoom (Abbott Laboratories Inc, Abbott Park, IL) (Fig. 36-2), have 4 D and 3.5 D adds at the IOL plane that yield approximate add powers of about 3.2 D and 2.8 D at the corneal plane, respectively. The optimum reading distance of any IOL model can be adjusted at the expense of the distance refraction. For a positively powered IOL, increasing the myopic distance refraction will strengthen the effect of the add power and shorten the reading

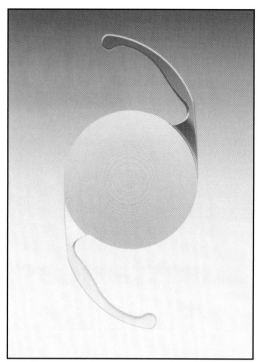

Figure 36-1. The Alcon Acrysof Restor multifocal IOL.

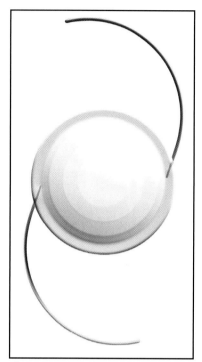

Figure 36-2. The AMO Rezoom multifocal IOL.

distance while a hyperopic distance refraction will do the opposite. Although modern multifocal IOLs provide some intermediate vision, it is adequate to consider only the near and far point of the postoperative eye for planning purposes.

In practical terms, the surgeon must select the IOL add power and distance refraction that represents the best compromise for each patient. In Table 36-1, the PreVize Multifocal IOL Planning Guide (Moran Research and Consulting, Inc, Houston, TX; PreVize Calculation Center, www.previze.com) shows the optimum reading distance and expected distance UCVA for several combinations of refractive target and IOL models within a clinically realistic range.

As an example, an avid golfer and sport fisherman who wishes to see the ball on the tee and also tie fishing line would likely be more content with a +4.0 D add and -0.75 D refraction than a +3.5 D add and a plano refraction. A seamstress who prefers a short working distance for intricate needle work and also enjoys bird watching would be expected to prefer the latter IOL and refraction.

Toric IOLs

The best possible result from a toric IOL implantation is achieved when the IOL and the principle meridians of the *postoperative* cornea are precisely aligned. While the actual power and astigmatism of the postoperative cornea are obviously unknown preoperatively, they can be approximated by adding the mean change in keratometry, determined for a group of previously operated patients, to the preoperative keratometry of the eye to

Table 36-1.

PreVize Multifocal IOL Planning Guide

IOL with +3.5 D Add (eg, AMO Rezoom)				IOL c +4 D Add (eg, Alcon AcrySof ReStor)			
Target Ref D (SEQ)	Near Point Inches	Far Point Feet	Distance UCVA*	Target Ref D (SEQ)	Near Point Inches	Far Point Feet	Distance UCVA*
1.00	22	N/A	20/40	1.00	18	N/A	20/40
0.75	19	N/A	20/30	0.75	16	N/A	20/30
0.50	17	N/A	20/25	0.50	15	N/A	20/25
0.25	15	N/A	20/20	0.25	13	N/A	20/20
0.00	14	Infinity	20/20	0.00	12	Infinity	20/20
-0.25	13	13.1	20/20	-0.25	11	13.1	20/20
-0.50	12	6.6	20/25	-0.50	11	6.6	20/25
-0.75	11	4.4	20/30	-0.75	10	4.4	20/30
-1.00	10	3.3	20/40	-1.00	9	3.3	20/40

*This approximation assumes <0.5 D cylinder and a 3 to 4 mm pupil.

be operated. This average keratometric change, known as the mean surgically induced astigmatism (SIA), can be easily calculated online.[1] Cylinder correction is highly dependent on the proper rotational positioning of the IOL relative to the cornea. A 10 degree misalignment will reduce the efficacy of the cylinder correction by 30%.

The refractive target should be chosen so as not to "flip the axis of astigmatism." Patients do not tolerate an inversion of the astigmatism axis they have grown accustomed to. In practice, it is best to choose an add power that leaves the patient with undercorrected rather than overcorrected cylinder.

Further Refinements

Most currently available software for calculating toric IOL power uses simple vector-based mathematics (AcrySof Toric IOL Web Based Calculators). These methods are limited to corneas with orthogonal principal meridians. Corneas with principle meridians that deviate from orthogonal by at least 10 degrees (present in about 10% of astigmats) cannot be analyzed by these methods.[1] More sophisticated ray tracing and matrix methods must be used to analyze these cases properly. A toric IOL power calculator that can correctly analyze nonorthogonal Ks is available online (PreVize Calculation Center, www.previze.com and PreVize Optimized IOL Power Calculation Web Service for the STAAR Toric IOL).

Finally, the ability to correct astigmatism is also limited by the reliance on corneal surface measurements, such as keratometry, to deduce the optical properties of the cornea. The shape and refractive power of the back surface of the cornea is only loosely related to the shape of the anterior surface. The relative alignment, tilt, and rotation of back surface torus with respect to the anterior surface, all of which effect the optical power of the cornea, varies from eye to eye. This will continue to add a random error to toric IOL power calculations until more accurate methods of measuring the optical properties of the cornea are available.

Key Points

- Use consistent biometric and surgical techniques.
- Use a modern IOL power formula and optimize IOL constants, which can be easily calculated using Hoffer Programs®, Holladay IOL Consultant®, or online (PreVize Calculation Center, www.previze.com).
- Match the distance refraction and add power of multifocal IOLs to the patients most important visual tasks.
- Determine and use personalized SIA in planning IOL procedures.
- Select toric add powers that do not over correct refractive cylinder.

Reference

1. Harris WF. Interpretating nonorthogonal keratometric measurements. *Ophthalmic Physiol Opt.* 2001;21(3):253–254.

Special Circumstances: Pediatric Eyes

Scott K. McClatchey, CAPT, MC, USN, MD

Background

There are two concerns in choosing an IOL power for a child: the initial IOL calculation and the growth of the eye. For children younger than about age 10, the growth of the eye has an overwhelming impact.

Formulas for calculating the initial IOL power for young children are relatively inaccurate, compared to the same formulas in adults. Andreo et al[1] studied 47 consecutive pseudophakic patients age 3 months to 16 years and found no significant difference in accuracy between the several IOL calculation formulas: the average initial postoperative refractive error was between 1.2 and 1.4 D for all formulas.

Gordon and Donzis[2] showed that a normal child's eye has little change in refraction (0.9 D from birth through adulthood) because the power of the natural lens decreases dramatically as the eye grows. McClatchey and Parks[3] showed that the refraction of aphakic children's eyes has a large myopic shift (10 D from infancy through adulthood). The mean refraction of these eyes follows a logarithmic curve from infancy through age 20. The optics of a growing pseudophakic eye result in a *magnification* of this aphakic myopic shift. The growth of a pseudophakic child's eye therefore results in a large myopic shift.

Table 37-1.

PREDICTION TABLE FOR *TYPICAL* PEDIATRIC PSEUDOPHAKIC EYES[1]

Age at surgery	IOL power[2] D	Initial PO Rx[2]	Predicted refractions at a given age[1]				
			1 yr.	2 yr.	4 yr.	8 yr.	20 yr.
3 mo.	26.9	+7.00	3.30	1.68	0.07	-1.54	-3.64
6 mo.	25.1	+6.50	4.23	2.13	0.04	-2.03	-4.73
1 yr	24.5	+5.00	5.00	2.77	0.70	-1.35	-4.02
2 yr	22.3	+4.00	NA	4.00	2.01	0.03	-2.53
3 yr	22.3	+3.00	NA	NA	2.17	0.19	-2.40
4 yr	22.0	+2.25	NA	NA	2.25	0.28	-2.29
6 yr	21.0	+1.50	NA	NA	NA	0.70	-1.82
8 yr	20.4	+1.00	NA	NA	NA	1.00	-1.50

Assumed A-constant = 118.0. Calculations are based on eyes with a normal RRG; variations in RRG and initial ocular measurements will significantly affect these predictions. The large variance in RRG will lead to a large range of ultimate refractions: these are the expected averages. These IOL power and initial postoperative refractions are for example only, and are not our recommendations.

Rate of Refractive Growth

Because aphakic refraction follows a logarithmic decline, a plot of refraction vs. the log of age for these eyes is a straight line. The slope of this line is defined as the "Rate of Refractive Growth" (RRG). RRG has units of diopters, and can be calculated for pseudophakic eyes by mathematically "removing" the IOL. There is little difference in the RRG between aphakic and pseudophakic eyes over age 6 months. Recent unpublished analysis indicates that there is no significant difference, ie, putting an IOL in a child's eye does not affect its refractive growth.

RRG is smaller in children who have surgery at less than 6 months of age. However, in children who have surgery after 6 months of age, RRG is not affected by age at surgery, type of cataract, initial refraction, or controlled glaucoma. No other studied factor has been found to consistently influence RRG. The value of RRG has a large standard deviation (ie, some eyes grow faster than others).

Discussion

RRG is useful in clinical practice because it allows prediction of refraction in pseudophakic eyes. It is also useful in research. I believe it is the best way to analyze refractive changes in pseudophakic children.[4] Calculating RRG eliminates the large confounding factors of non-linear growth of children's eyes and the variables of age at surgery, length of follow-up and variations in IOL power (Table 37-1).

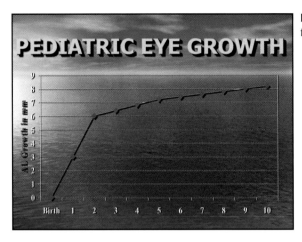

Figure 37-1. Graph of axial length growth from birth to age 10.

Note From the Editor

Another consideration is decreasing the power of the primary IOL by the difference that will occur with aging to maturity. The graph of eye growth in infancy and childhood (Fig. 37-1) can be helpful in this regard by projecting the change from implantation to adulthood. That subtracted power can be implanted as a second piggyback IOL over the top of the primary or as a phakic IOL, either of which could be removed when the patient grows up. In the extremely young, this could be done using three lenses.

References

1. Andreo LK, Wilson ME, Saunders RA. Predictive value of regression and theoretical IOL formulas in pediatric intraocular lens implantation. *J Pediatr Ophthalmol Strabismus.* 1997;34(4):240–243.
2. Gordon RA, Donzis PB. Refractive development of the human eye. *Arch Ophthalmol.* 1985;103(6):785–789.
3. McClatchey SK, Parks MM. Myopic shift after cataract removal in childhood. *J Pediatr Ophthalmol Strabismus.* 1997;34(2):88–95.
4. McClatchey SK, Hofmeister EM. Intraocular lens power calculation for children. In: Wilson ME, Trivedi RH, Pandey SK, eds. *Pediatric Cataract Surgery: Techniques, Complications and Management.* Philadelphia, PA: Lippincott, Williams & Wilkins; 2005:30–37.

The views expressed in this chapter are those of the author and do not necessarily reflect the official policy or position of the Department of the Navy, Department of Defense, or the United States Government.

Special Circumstances: Piggyback IOLs

Kenneth J. Hoffer, MD

Gayton[1] was the first to perform and report implanting a piggyback IOL. A piggyback lens is defined as an IOL that is inserted on top of another IOL. They may be placed together in the capsular bag directly in contact with each other, in the ciliary sulcus away from the other IOL, or in the anterior chamber. Piggyback lenses can be either placed primarily (with the first IOL) or secondarily over a previously implanted IOL. The need for piggyback lenses arises in extremely high hyperopes (primary) and in cases of postoperative IOL power error (secondary).

Primary Piggyback Calculation

If both IOLs are placed within the capsular bag, it is estimated that the anterior IOL forces the posterior IOL more posteriorly, a distance equal to half the central thickness of the anterior lens. This was shown by Baumeister and Kohnen[2] using Scheimpflug studies in 2006. They showed this did not occur if the piggyback lens was placed in the ciliary sulcus. The shift causes the posterior lens (whose focal point is moved more posteriorly) to require more power to maintain the same focus. This effect diminishes the thinner (lower power) the anterior lens is, and a thinner lens is easier to remove if that should be necessary later. Primary piggyback lenses need special calculations to adjust for this posterior lens shift. This can be done by simply adding one-half the central thickness of the anterior IOL to the ELP being used by the formula to calculate the power of the posterior IOL.

Here is an example of how this would be done. If we have an eye that requires a 36 D IOL power and we have to split the power between two lenses, the highest power we can obtain for a posterior chamber (PC) lens is 30 D. Since we know we will need more than 36 D in this situation, we obtain from the manufacturer the central thickness of a 7 D IOL, which is 0.942 mm. We now add one-half of this thickness (0.47 mm) to the ELP the formula calculates for the posterior lens (eg, 4.37 mm) to obtain the ELP we need to use (4.37 + 0.47 = 4.84.) We should now use 4.84 as the ELP to perform the final calculation. The Holladay IOL Consultant® has this calculation built in to the program, making this easier to perform.

Secondary Piggyback Calculation

Secondary lenses can be calculated using the refraction formula of Holladay[3] (used for ahakic eyes and phakic IOL power calculation) or by a more simple formulation based on the fact that the healed primary IOL is more stable. Due to the different effect on vertex power changes between plus and minus lenses, the following formulation works well:

$$\text{Hyperopic Error: Piggyback IOL} = 1.5 \times Rx_{Error}$$

$$\text{Myopic Error: Piggyback IOL} = 1.0 \times Rx_{Error}$$

where Rx_{Error} = the postoperative spherical equivalent refractive error needed to be corrected.

Habot-Wilner and associates[4] from Israel reported excellent prediction results using a slight variation of the formulation for hyperopic error:

$$\text{Hyperopic Error: Piggyback IOL} = 1 + 1.4 \times Rx_{Error}$$

They reported a mean prediction error on 10 eyes of 0.46 ±0.40 D.

Shammas[5] has suggested a formulation utilizing the A constant of the IOL as follows:

$$\text{Hyperopic Error: Piggyback IOL} = [Rx_{Error}/[(0.03 * (138.3 - A)] - 0.5$$

$$\text{Myopic Error: Piggyback IOL} = [Rx_{Error}/[(0.04 * (138.3 - A)] - 0.5$$

References

1. Gayton JL, Sanders V, Van Der Karr M, Raanan MG. Piggybacking intraocular implants to correct pseudophakic refractive error. *Ophthalmology.* 1999;106(1):56–59.
2. Baumeister M, Kohnen T. Scheimpflug measurement of intraocular lens position after piggyback implantation of foldable intraocular lenses in eyes with high hyperopia. *J Cataract Refract Surg.* 2006;32:2098–2104.
3. Holladay JT. Refractive power calculations for intraocular lenses in the phakic eye. *Am J Ophthalmol.* 1993;116(1):63–66.
4. Habot-Wilner Z, Sachs D, Cahane M, et al. Refractive results with secondary piggyback implantation to correct pseudophakic refractive errors. *J Cataract Refract Surg.* 2005;31(11):2101–2103.
5. Shammas HJ. The Shammas refractive equations. In: H. Shammas, ed. *IOL Power Calculations.* Thorofare, NJ: SLACK Incorporated; 2004:60-61.

Special Circumstances: Silicone Oil Power

Kenneth J. Hoffer, MD

In calculating IOL power, there are 2 major problems when the posterior segment is filled with silicone oil:

1. A *physical* one affecting ultrasound AL measurement due to the slow US velocity through the oil filling the large length of the eye.
2. An *optical* one affecting the overall optical power of the eye due to the refractive index of the oil compared to vitreous.

Often, such eyes have serious retinal conditions which preclude a postoperative acuity of 20/40 or better and thus postoperative studies are quite difficult.

Silicone Oil Ultrasound Effect

Silicone oil-filled eyes are an especially vexing problem because the ultrasound wave is so slowed down crossing the posterior segment that it is often impossible to get a reading at all. The sound velocity drops from the vitreous velocity of 1532 m/sec to the silicone oil velocity of 980 m/sec; a 36% decrease. There are also differences in velocity depending upon the type of silicone oil. Preset ultrasound instruments that do not allow changing the velocities makes it impossible to obtain an accurate reading. It is also difficult in some eyes to determine what percentage of the vitreous body is filled and depending on the position of the patient, what parts the beam is going through. Some have suggested multiplying the AL reading obtained with ultrasound by 0.71. I have no experience with this suggestion.

Silicone oil eyes are best measured with the IOLMaster or the LenStar because the optical laser is not appreciably affected by the silicone oil, though fixation may be a problem.

This problem could easily be solved if all surgeons replacing the vitreous with silicone oil would ensure that an immersion A-scan or IOLMaster AL measurement was taken and recorded prior to doing so. The result should be given to the patient with instructions to keep it safe and provide it to any surgeon planning to perform cataract surgery.

Other alternatives are not performing a primary implant and later performing a secondary IOL after the aphakic refraction is obtained. Also, one could consider a piggyback lens or phakic IOL to correct the error.

Silicone Oil Refractive Effect

The second problem that arises when the vitreous is replaced with silicone oil is that the refractive index of the oil is much less than that of the vitreous and the silicone oil acts as a negative-powered lens in the eye. This must be offset with more power in the IOL. Obviously, this factor can be ignored if the silicone oil will be removed completely at some time in the future.

This optical effect is heavily dependent upon the shape factor of the back surface of the IOL, such that a meniscus lens with the concave surface facing posteriorly causes practically no effect. But these IOLs are no longer commercially available. With a plano-convex lens, with the plano surface facing posteriorly, causes a moderate effect such that 2 to 3 D must be added to the IOL power to compensate for this silicone effect.

The greatest problem is with biconvex IOLs that have a convex surface facing the silicone oil. This situation requires 3 to 5 D of added IOL power to offset this effect.

Special Circumstances: Effect of IOL Tilt on Astigmatism

Susana Marcos, PhD, FOSA, FEOS
with Patricia Rosales, PhD; Alberto de Castro, MSc; and Ignacio Jiménez-Alfaro, MD, PhD

Introduction

As IOLs become more sophisticated, the question arises whether the potential improvement in ocular optical quality of the new designs might be compromised by proper IOL centration.[1] The use of aberrometry allows us to measure the optical aberrations of the eye including astigmatism and, in particular, to evaluate optical quality after cataract surgery in pseudophakic patients. However, additional new tools allow full evaluation of the different contributions to optical degradation (astigmatism, in particular) in eyes with IOLs. Astigmatism in pseudophakic eyes can arise from various sources, including natural corneal astigmatism, astigmatism induced by the incision, the eccentric fixation of the fovea, or tilt and decentration of the IOL. Among them, the incision-induced astigmatism is the most relevant.[2-4]

Whether normal amounts of tilt and decentration of the IOL can cause significant amounts of astigmatism (and other higher order aberrations) is of high interest. It is particularly important in new aspheric designs, where the correction of spherical aberration (and improved retinal image quality) is aimed. Despite initial concerns that aspheric IOLs may be more susceptible to increasing higher order aberrations (such as coma),[1] recent studies show that this does not seem to be the case in comparison with spherical IOLs.[5-8] Future IOL trends aiming at individual customization of designs also rely on the absence of relevant induced IOL tilt and decentration.[9]

This chapter will present simple computations on the theoretical astigmatism induced by a tilted lens. However, it will be shown that real measurements are essential to assess the impact of real amounts of tilt and decentration on astigmatism. Custom-developed instrumentation to measure IOL tilt and decentration will be described. In addition, a custom model eye will be presented that allows accurate estimations of the astigmatism induced by IOL tilt and decentration (in comparison with a perfectly centered IOL), and in relation to other sources of astigmatism, in particular, and optical degradation in general.

Theoretical Astigmatism Induced by a Tilted Lens

Simple aberration theory shows that a narrow beam of light entering a spherical lens obliquely (not parallel to the axis of the lens), will show *marginal oblique astigmatism*. In general, a horizontal or vertical tilt of the lens potentially induces an astigmatic effect. Although differences are found with the geometry of the lens, simple estimations can be performed for a thin lens. Theoretically, a thin lens of power P in air, located in the pupil plane and tilted by angle α, generates an astigmatism of A with an angle perpendicular to the tilt (α) axis given by[10]

(1) $$A = P \cdot [1 + (\sin \alpha)^2 / 3] * (\tan \alpha)^2$$

where A = astigmatism (D), P = power of IOL (D)

For example, an ophthalmic glass in air of +22 D, tilted 10° would produce an oblique astigmatism of 0.69 D.

$$A = 22 \cdot [1 + (\sin 10°)^2/3] * (\tan 10°)^2 = 22 \cdot [1 + (0.1736)^2/3] * (0.1763)^2$$

$$A = 22 \cdot [1.01005] * (0.03109) = 22(0.0314) = 0.6908 \text{ D}$$

Although indicative of the order of magnitude of the astigmatism induced by a tilted IOL, equation (1) is largely simplified to be applied in pseudophakic eyes, where:
a. the IOL is immersed in aqueous
b. the rays of light converge on the IOL from the cornea
c. the cornea and IOL form a compound optical system
d. the IOL is not a thin lens, and its design has an impact on optical quality, including the degradation caused by off-axis viewing
e. the IOL does not lie on the pupil plane of the system
f. the eye is not a centered optical system, with the fovea tilted with respect to the "optical axis"
g. the IOL is both tilted and decentered with respect to the pupillary axis
h. several factors, including corneal astigmatism and incision-induced astigmatism, contribute to the total astigmatism in the eye
i. the actual amount and orientation of tilt and decentration of the IOL should be considered for correct estimates of their impact on image quality

Measurement of IOL Tilt

IOL tilt and decentration can be measured in vivo using Purkinje imaging (Fig. 40-1A),[11-14] and Scheimpflug imaging (Fig. 40-1B).[15-16] Both techniques have been validated

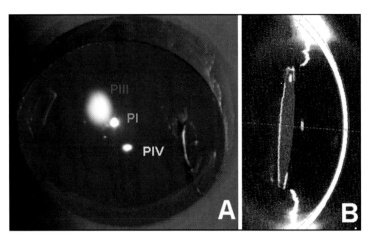

Figure 40-1. Examples of images obtained from Purkinje imaging (A) and Scheimpflug imaging (B) in pseudophakic eyes. The reflections are PI (anterior cornea,) PIII (anterior IOL surface) and PIV (posterior IOL surface.) The surface images in the Scheimpflug photo have been fitted by conics.

using physical eye models and used to estimate tilt and decentration of both the natural lens in phakic eyes and IOLs in patients.

Purkinje images are reflections from the ocular surfaces of the eye. A widespread implementation of a Purkinje imaging system to measure tilt and decentration assumes that the locations of the Purkinje images of a point source are linearly related with eye rotation (β), IOL tilt (α), and IOL decentration (d).[11,17,18]

$$PI = A\beta$$

$$PIII = B\beta + C\alpha + Dd$$

$$PIV = E\beta + F\alpha + Gd$$

where PI, PIII and PIV are the locations of the Purkinje reflections from the anterior cornea, anterior and posterior lens respectively, and A-G are coefficients customized to the anatomical parameters of the eye.

Scheimpflug imaging allows capture of anterior segment images with a large depth of focus, although they are subject to optical and geometrical distortion.[19] The distortion-corrected Scheimpflug images can be processed to obtain the pupillary axis (joining the center of rotation of the cornea and the pupil center) and the IOL axis (joining the center of curvature of the anterior and posterior IOL surfaces), and therefore the tilt of the IOL.

Fig. 40-1 shows typical images (Purkinje A, Scheimpflug B) obtained in experimental systems on pseudophakic eyes from the Visual Optics and Biophotonics Laboratory in Madrid, Spain. Fig. 40-2 shows the amounts of tilt measured using Purkinje imaging in 30 pseudophakic eyes. The average tilt of the IOL was 0.25° (around the horizontal axis) and 1.79° (around the vertical axis). The average amount of decentration of the IOL (not depicted) was 0.237 mm horizontally and -0.039 mm vertically. Tilt and decentration tended to be mirror-symmetric in right and left eyes.

Customized Pseudophakic Computer Eye Models

The impact of IOL tilt (and decentration) can be accurately evaluated using customized computer eye models.[7,20] These models are built using the anatomical parameters measured for an eye. The use of these models overcomes all the limitations of the theoretical simple computations (Formula 1, previously) and allows understanding the contribution

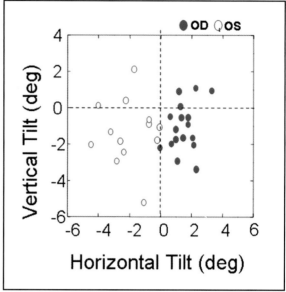

Figure 40-2. IOL tilt measured using Purkinje images in 30 eyes (OD and OS). Horizontal tilts stands for tilt around the horizontal axis and vertical tilt is for tilt around the vertical axis.

of each factor to the optical quality (including astigmatism) in pseudophakic eyes, and most importantly, to perform this evaluation on an individual basis.

We have presented customized eye models[7] that include anterior surface topography (measured with videokeratoscopy), anterior chamber depth and axial length (measured with low coherence interferometry), IOL tilt and decentration and foveal misalignment (using Purkinje imaging), and IOL geometry. The posterior corneal surface (unmeasured) was assumed spherical.

Fig. 40-3 shows a schematic diagram of the pseudophakic eye model used to evaluate the impact of the different factors (including IOL tilt) on the optical aberrations (including astigmatism), along with the various sources of the measured anatomical parameters.

Effect of Real IOL Tilt on Ocular Astigmatism

Computer eye models allow prediction of the ocular aberrations by ray tracing. These estimates can be compared to actual measurements of aberrations on the same eyes. We computed the aberrations in 12 customized pseudophakic eye models implanted with aspheric IOLs (Acrysof IQ) and compared them to aberrations measured on those patients.[7] The evaluation of the impact of IOL tilt and decentration on the aberrations was achieved by comparing the optical quality, assuming that the IOL was centered on the pupillary axis (no IOL tilt and decentration), or subject to the actual measured combinations of tilt and decentration.

Predicted and measured aberrations are described by sets of Zernike coefficients (up to the 7th order.) For the purposes of this chapter, we analyzed the astigmatism coefficients in the Zernike expansion (astigmatism at 0/90° Z^2_2 and astigmatism at 45° Z^{2}_{2}).

Fig. 40-4 compares corneal astigmatism, measured total astigmatism, and simulated astigmatism (assuming no IOL tilt/decentration and with their real amounts). We found a good correlation between the simulated and measured astigmatism. There is a consistent

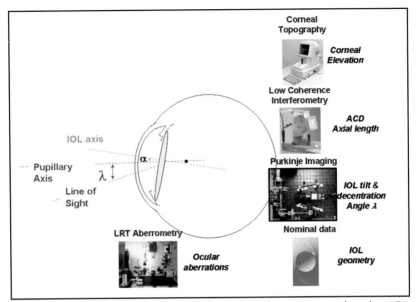

Figure 40-3. Computer eye model showing the relevant axes and angles (IOL tilt α and foveal misalignment λ,) and the instruments used to customize it for each eye. Estimated aberrations (including astigmatism) by ray tracing on this eye model were compared to the aberrations measured using Laser Ray Tracing Aberrometry.

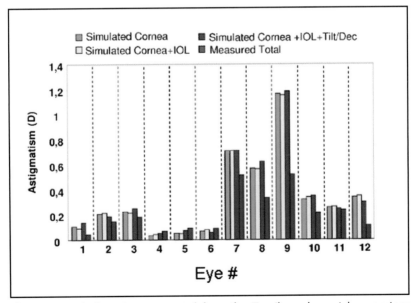

Figure 40-4. Astigmatism computed from the Zernike polynomial expansions in simulated and measured wave aberrations in 12 pseudophakic eyes showing: anterior corneal astigmatism (green); simulated astigmatism from computer eye models; a) assuming a centered IOL (yellow), and b) assuming a tilted/decentered IOL (red) as well as measured astigmatism (blue).

overestimation of the simulated astigmatism by 17.3% for Z^2_2 and by 8.7% for Z^2_{-2} with respect to the real measurements. This result is indicative of a compensatory role of the astigmatism by the real posterior corneal surface.[21-22] The results for the cornea alone are similar to those of the model with the IOL (either lined up or tilted/decentered), indicating that the cornea is the major source of astigmatism in pseudophakic eyes and IOL tilt and decentration play a negligible role in the postoperative astigmatism.

Conclusions

1. IOL tilt can be measured accurately using imaging techniques, provided with validated routines for image analysis and quantification.
2. With state-of-the-art monofocal IOLs, tilt is typically less than 5°.
3. Computer eye models can be used to accurately estimate the contribution of IOL tilt and decentration on optical quality. These models consider the individual interactions among all geometrical parameters of the eye.
4. In modern cataract surgery with monofocal IOLs, the effect of IOL tilt on ocular astigmatism is neglible.

References

1. Atchison DA. Design of aspheric intraocular lenses. *Ophthalmic Physiol Opt.* 1990;11(2):137-146.
2. Guirao A, Tejedor J, Artal P. Corneal aberrations before and after small-incision cataract surgery. *Invest Ophthalmol Vis Sci.* 2004;45:4312-4319.
3. Marcos S, Rosales P, Llorente L, Jimenez-Alfaro I. Change of corneal aberrations after cataract surgery with two types of aspheric intraocular lenses. *J Cataract Refract Surg.* 2007;33:217-226.
4. Jacobs BJ, Gaynes BI, Deutsch TA. Refractive astigmatism after oblique clear corneal phacoemulsification cataract incision. *J Cataract Refract Surg.* 1999;25:949–952.
5. Barbero S, Marcos S, Jimenez-Alfaro I. Optical aberrations of intraocular lenses measured in vivo and in vitro. *J Optom Soc.* 2003;20:1841-1851.
6. Marcos S, Barbero S, Jiménez-Alfaro I. Optical quality and depth-of-field of eyes implanted with spherical and aspheric intraocular lenses. *J Refract Surg.* 2005;21:223-235.
7. Rosales P, Marcos S. Customized computer models of eyes with intraocular lenses. *Optics Express.* 2007;15:2204-2218.
8. Marcos S, Rosales P, Llorente L, Barbero S, Jiménez-Alfaro I. Balance of corneal horizontal coma by internal optics in eyes with intraocular artificial lenses: Evidence of a passive mechanism. *Vision Res.* 2008;48:70-79.
9. Barbero S, Marcos S. Analytical tools for customized design of monofocal intraocular lenses. *Optics Express.* 2007;15:8576-8591.
10. Fannin T, Grosvenor TP (Eds). *Clinical Optics* (2nd edition). New York, NY: Elsevier, Inc.;1996:48-50.
11. Rosales P, Marcos S. Phakometry and lens tilt and decentration using a custom-developed Purkinje imaging apparatus: Validation and measurements. *J Opt Soc Am A.* 2006;23:509-520.
12. Tabernero J, Benito A, Nourrit V, Artal P. Instrument for measuring the misalignments of ocular surfaces. *Optics Express.* 2006;14:10945-10956.

13. Schaeffel F. Binocular lens tilt and decentration measurements in healthy subjects with phakic eyes. *Invest Ophthalmol Vis Sci.* 2008;49:2216-2222.
14. Rosales P, de Castro A, Jiménez-Alfaro I, Marcos S. Intraocular lens alignment from Purkinje and Scheimpflug imaging. *Clinical and Experimental Optometry.* 2010;93;400-408.
15. de Castro A, Rosales P, Marcos S. Tilt and decentration of intraocular lenses in vivo from Purkinje and Scheimpflug imaging - Validation study. *J Cataract Refract Surg.* 2007;33:418-429.
16. Coppens JE, van der Berg TJ, Budo CJ. Biometry of phakic intraocular lens using Scheimpflug photography. *J Cataract Refract Surg.* 2005;31(10):1904-1914.
17. Phillips P, Perez-Emmanuelli J, Rosskothen HD, Koester CJ. Measurement of intraocular lens decentration and tilt in vivo. *J Cataract Refract Surg.* 1988;14:129-135.
18. Barry JC, Dunne M, Kirschkamp T. Phakometric measurement of ocular surface radius of curvature and alignment: Evaluation of method with physical model eyes. *Ophthalmic Physiol Opt.* 2001;21:450-460.
19. Rosales P, Marcos S. Pentacam Scheimpflug quantitative imaging of the crystalline lens and intraocular lens. *J Refract Surg.* 2009;25:421-428.
20. Tabernero J, Piers P, Benito A, Redondo M, Artal P. Predicting the optical performance of eyes implanted with IOLs to correct spherical aberration. *Invest Ophthalmol Vis Sci.* 2006;47:4651-4658.
21. Dunne M, Royston J, Barnes D. Posterior corneal surface toricity and total corneal astigmatism. *Optom Vis Sci.* 1991;68:708-710.
22. Dubbelman M, Sicam V, Van der Heijde GL. The shape of the anterior and posterior surface of the aging human cornea. *Vision Research.* 2006;46:993-1001.

Special Circumstances: Aniseikonia and Anisometropia

Kenneth J. Hoffer, MD

Anisometropia is defined as a difference in refractive error between the two eyes, and is often confused with aniseikonia. *Aniseikonia* is defined as the binocular status of unequal image sizes projected on the two maculas (*iseikonia* means equal images). Aniseikonia can lead to a discomforting annoyance termed *asthenopia* if the percentage difference is greater than 14%. Obviously, in young children under 6 years of age this situation can lead to *amblyopia*.

The causes of aniseikonia are due to a combination of axial length, corneal power, lens power, and spectacle correction. Because of various biological differences in the above parameters, it is possible for an individual to have anisometropia but still be iseikonic as well as the opposite being true.

Hermann Gernet of Münster (now living in Würzburg), Germany, devoted a lot of time in the late 1960s and early 1970s to the evaluation and prevention of this condition, especially after IOL implantation. He used complex formulas and mainframe computers to perform the calculations. When I visited him in 1974, it became obvious that his calculations would be a very complex method to use as a routine for the average surgeons beginning lens implantation in the US at that time. Following Gernet's lead, Colenbrander[1] included a specific formula to calculate an aniseikonic IOL power in his landmark paper of 1973.

In 1974, I[2] began my work in this field using the Colenbrander formula and performed a comparison of the emmetropic and aniseikonic formulas to calculate IOL powers. The result yielded an eye that would be approximately 1 to 1.50 D more myopic than the target refraction originally aimed for. Since my goal was emmetropia and not purposely causing myopia, I refrained from using the aniseikonia formula. Over the next several years, there did not seem to be any patient complaints attributable to aniseikonia. This is the reason you do not hear this subject being discussed regarding IOL power. On the other hand, perhaps some of the vague complaints patients relay after IOL implantation might be attributable to this problem. More research is needed.

References

1. Colenbrander MC. Calculation of the power of an iris clip lens for distant vision. *Br J Ophthalmol.* 1973;57(10):735–740.
2. Hoffer KJ. Intraocular lens calculation: The problem of the short eye. *Ophthalmic Surg.* 1981;12(4):269–272.

Preventing IOL Power Errors

Kenneth J. Hoffer, MD

Table 42-1 gives a summarized outline of the many things that can be done in clinical practice to improve IOL power calculation accuracy and prevent the dreaded IOL power surprise.

Prevention of Common IOL Power Errors

- Know more about IOL power calculation than your employees.
- Employ a well-trained, experienced technician.
- Use only the IOLMaster, Haag-Streit LENSTAR LS900, or immersion A-scan to measure the AL.
- Carefully evaluate the IOLMaster scan for reliability.
- If importing external K readings, be sure to set the index of refraction (IR) to 1.3375 in the setup screen of the IOLMaster for the Hoffer Q formula to yield correct results.
- Suspect a staphyloma in eyes >25 mm: Use IOLMaster and/or Shammas A/B-scan technique.
- Use CALF AL method: measure the eye using 1532 m/s and add +0.32 mm to the result to correct for any error in sound velocity (useful in very long and short eyes).
- Silicone oil eyes need IOLMaster if possible or ultrasound AL times 0.71.

Table 42-1.

SUMMARY FOR PREVENTING IOL POWER ERRORS

Summary

- Calculate power accurately
- Order PC and AC powers
- Fill out OR sheet
- Place OR sheets on wall and scope
- Prior to insertion, check power against sheet

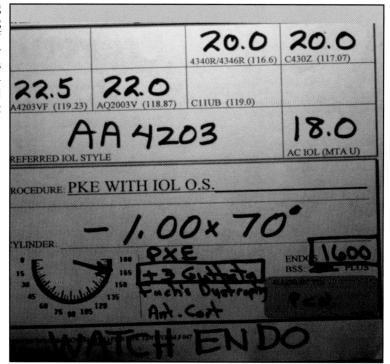

Figure 42-1. Operating room IOL sheet listing the various powers of potential IOLs that may be needed, as well as amount and axis of cylinder and the endothelial cell count. Note *red* sheet for *right* eyes.

- Regularly calibrate manual keratometers.
- Keep the CL out completely for 2 weeks prior to keratometry (at least in one eye).
- Use the Hoffer Q formula in eyes <22 mm and in post-refractive surgery short eyes.
- Use the Holladay 1 formula in eyes 24.5 to 26 mm in length.
- Use the SRK/T formula in eyes longer than 26 mm.
- *Never use the SRK Regression formulas (SRK I or II).*
- Personalize your ELP factors in the formulas.
- The surgeon should personally select the IOL power for the individual patient.
- Prepare a sheet with all IOL powers that may be needed (Fig. 42-1) and place it on the wall in the OR for the staff (Fig. 42-2) and a minimized copy on the microscope

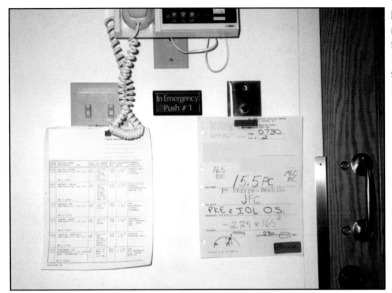

Figure 42-2. Operating room IOL sheet attached full size to the wall, readily available to the nursing staff. Note Yellow sheet for Left eyes.

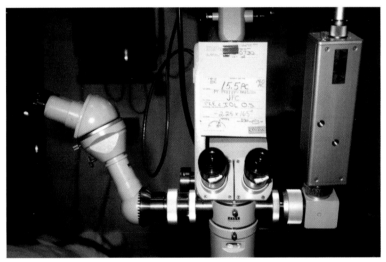

Figure 42-3. Operating room IOL sheet attached as a minimized copy to the operating microscope, readily available to the surgeon while operating.

for the surgeon to verify the correct IOL power (Fig. 42-3). Use red paper for right eyes and yellow paper for left eyes. Other important data can also be added to the sheet.

- Use the various methods available to better calculate an IOL power in post-refractive surgery corneas:
  - Clinical History Method if pre-LASIK refraction and K are available.
  - Contact Lens Method if eye can be refracted effectively. Have a set of PMMA hard CLs in the clinic and be sure the staff knows where they are.
  - Shammas "No History" Formula: $K = 1.14 * K_{PO} - 6.8$.
  - Maloney or Koch Corneal topography methods.

- Use the Aramberri Double-K: Calculate the ELP using the preoperative K and the IOL power using the PO K.
- Speicher/Seitz, Savini, and Masket Methods.
- Haigis-L formula.
• Consider delaying the IOL implantation until the cornea has healed after a penetrating keratoplasty, rather than performing a "triple procedure."

Diagnosing and Treating IOL Power Errors

Kenneth J. Hoffer, MD

This chapter will discuss ways to diagnose IOL power errors early and treat them quickly for the benefit of the patient as well as the surgeon.

Diagnosing IOL Power Surprises

The biggest thing one can do to prevent medicolegal problems with IOL power error is to make it a routine to perform a K reading and full manifest refraction on the first postoperative day. With today's surgery, this is possible in 95% of cases and takes no longer to do than it does a week or two later. It has been so ingrained in surgeon's practices to just pinhole at one day PO and carefully refract much later.

Pinhole vision testing can miss power errors and delay diagnosis and treatment for several weeks. Day 1 refraction allows you to discover the problem early enough to take the patient back to the OR and correct the problem in the first 12 to 48 hours. The surgery will be less traumatic (than later) because the incision will be easy to reopen, the capsule has not scarred and healed around the IOL so it is much easier to remove and replace a new lens. The patient is immediately pleased with the improved uncorrected visual acuity (UCVA) and usually forgets that they had two procedures performed when asked months later. With excellent UCVA (thus no damages), the opportunities for medicolegal actions are almost completely eliminated.

On the other hand, when weeks go by before the diagnosis is made and then several months pass as the surgeon attempts to mollify the patient's concerns, the patient often

Figure 43-1. McReynolds IOL power analyzer.

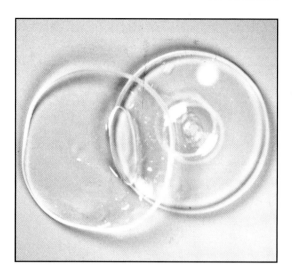

is recommended to see another ophthalmologist by their friends and relatives, unbeknownst to the surgeon. With an obvious IOL power error, it is difficult for the consulting surgeon to refrain from using the words, "…the wrong lens was inserted." That one word "wrong" becomes misinterpreted by the patient as "the surgeon made an error." All this can be prevented by early diagnosis and treatment. The patient is the first to benefit by having the problem solved in <48 hours.

Treating IOL Power Surprises

- In trying to determine the cause of the IOL power error, it is customary to re-measure the AL of the newly pseudophakic eye. If using ulrtrasound, this necessitates the use of the proper ultrasound velocity depending on the material of the IOL as described in Chapter 4. It is also important to ignore the sometimes large US spikes that appear in the mid-vitreous which are reduplication echoes caused by the IOL.
- It obtaining a remeasure of the AL, it might be wise to ask that it be performed in another colleague's office in case there may be a problem with your equipment. It is also wise if that colleague is not considered a close friend of yours.
- The best option is to have the AL remeasured using an IOLMaster, if that has not already been done. Be sure to adjust the menu to the appropriate pseudophakic eye type.
- If the eye has already healed and the capsular bag is tightly scarred around the IOL, consider the use of a piggyback IOL in the ciliary sulcus or a phakic IOL. This is appropriate when the eye has healed beautifully and removal of the errant IOL would be more traumatic to the eye. For myopic error use 1 times the error and for hyperopic errors use 1.5 times the error.
- In those patients where repeat intraocular surgery is not possible or advisable, consider a minimal corneal refractive surgery.
- If one is interested in measuring the power of a removed IOL in the OR, there is a device that can do that. It is called the McReynolds Analyzer (Fig. 43-1), and can

Figure 43-2. IOL Power Club founding members and Executive Committee. *Front row left to right:* Jaime Aramberri, MD, San Sebastian, Spain; Kenneth J. Hoffer, MD, Santa Monica, CA; and Thomas Olsen, MD, Århus, Denmark. *Back row left to right:* Sverker Norrby, PhD, Groningen, Holland; H. John Shammas, MD, Lynwood, CA; and Wolfgang Haigis, PhD, Würzburg, Germany.

only be obtained from William McReynolds, MD. (He can be contacted at 217-222-6656.) It is also possible to request the manufacturer of the removed IOL to bring equipment to your OR to perform the measurement.

Conclusion

This is the next to last chapter in our IOL power book and we all hope you have found the details we have delved into useful in your clinical practice to improve your ability to calculate the IOL power your patients need. I wish to personally thank all the members of the IOL Power Club who have taken time to participate in this large endeavor. The Club was founded in 2005 by a group of surgeons and engineers (Fig. 43-2) to promote the science of this field (see Appendix A).

Simple steps and attention to detail can be very useful in preventing IOL power errors. Since performing the first American ultrasound IOL power calculation[1] in 1974, the past 35 years have seen great improvement in the accuracy of postoperative refractive prediction. Future improvements may someday eliminate the problems we have left.

Reference

1. Hoffer KJ. The history of IOL power calculation in North America. In: Kwitko ML, Kelman CD, eds. *The History of Modern Cataract Surgery.* The Hague, Netherlands: Kuglen Publications; 1998:193–20.

Future Directions in IOL Power Calculation: Intraoperative Refractive Biometry

Tsontcho [Sean] Ianchulev, MD, MPH

Success of cataract surgery is inherently linked to the accuracy of IOL calculation. As cataract surgery has evolved over the years, so have conventional methods of IOL power calculation. Ever since Dr. Fyodorov described the first IOL power formula in 1967 based on preoperative keratometry and axial length, a long array of theoretical and empiric formulas have lead to significant incremental improvements in IOL power calculation. Colenbrander (1973), Hoffer (1974), Binkhorst (1975), Van der Heijde (1975), SRK I (1980), Holladay 1 (1988), SRK II (1990), Hoffer Q (1993), Olsen (1995), Holladay 2 (1996), and Haigis (2000) are the more salient embodiments of continuous innovation in the field of IOL biometry. Today, the 3rd and 4th generation formulas are able to predict the final emmetropic power in more than 90% of standard cataract cases within ±1 D. As a result of this, as well as the concurrent advancements in phacoemulsification technology, modern cataract surgical intervention has moved to refractive lens exchange.

Despite significant differences across the various IOL formulas, they all share the same basic principle deriving from Fyodorov's original equation—they are based on preoperative anatomic parameters, such as axial length and corneal curvature. In essence, they derive an optical variable (diopters) from nonrefractive, anatomic ocular parameters. This has been greatly enabled by the advances in ultrasound and optical biometry and keratometry. It also successfully bypasses any confounding effect the cataractous lens would have on any preoperative refractive measurements, if such were to be used as a surrogate estimate of lenticular corrective power.

Today our patients have a much greater expectation of achieving their desired postoperative refraction, but some traditional challenges remain and some new ones have emerged. Conventional preoperative nonrefractive biometric approaches continue to struggle with the difficulties associated with very short and very long eyes, where most formulas see an appreciable attrition of effectiveness and precision. More significantly, the conventional approaches fail to deliver the same superior efficacy in cases after previous LASIK and PRK surgery. The laser ablation of the cornea complicates proper estimation of keratometric values which are at the core of all conventional formulas. For a detailed description of these challenges and potential mitigation strategies and solutions please refer to earlier chapters in this book.

Intraoperative Refractometry: Initial Concept and Methodology

One alternative methodology for IOL calculation offers a rather different approach to IOL calculation, which may present potential opportunities to address some of the remaining needs in the field. Intraoperative automated refractive biometry can be used to obtain the aphakic spherical equivalent of the eye right after the extraction of the cataractous lens. In this transiently-aphakic state (after lens extraction and before IOL implantation,) one can take a "refractive" biopsy of the eye with an autorefracting device (autoretinoscope, wavefront aberrometer, etc.) to measure the lower-order aberrations (sphere, cylinder) which can then provide the aphakic spherical equivalent. Assuming minimal distortion of ocular optics during surgery (as is typical of today's minimally invasive phaco techniques) as well as high accuracy of autorefracting devices, the aphakic spherical equivalent informs us about the refractive deficit of the aphakic eye at the vertex distance of measurement. Converting or correlating this to the power at the intraocular plane of final lens position can provide the necessary emmetropic IOL power. Theoretical analyses based on Bennett–Rabbetts[1] schematic eye variants demonstrate that the expected ratio between the aphakic spherical equivalent and the final emmetropic power is in the range of 1.75 to 2.01.

This new methodology is compelling in a number of ways. Because of its purely refractive approach, one would predict little confounding by the effect of prior corneal alteration from refractive surgery. AL and keratometry are not needed, as their resultant optical effect should be factored into the aphakic autorefraction. In fact, preoperative lens calculations would be eliminated in this intraoperative refractive paradigm where diagnostic biometry is done "on the table" after the lens extraction. And the surgical effect of the phaco incision on the cornea would also be captured in this intraoperative setting.

While aphakic refraction has been described previously as a useful alternative for IOL power calculation during secondary IOL implantation, applying the technique intraoperatively is not trivial. Initial clinical experience has shown that in order to achieve the full potential of this method, control over and experience with a number of variables is important. A reliable autorefractor such as the portable Retinomax (Nikon, Kanagawa, Japan) or the Nidek AR-20 device (Nidek, Co. Ltd, Hiroishi, Japan) should be used (Fig. 44-1A), but ideally it would be better with a surgical microscope-integrated device (Fig. 44-1B) such as the ORange (WaveTec Vision, Aliso Viejo, CA). Manual refraction is discouraged due to operator dependence and reliability. Vertex distance, visual axis centration, and parallax are important, as are post-phaco corneal status, intraocular pressure (over/under-filled AC), and type of viscoelastic chamber maintainer. Fortunately, with the advent of new intraoperative autorefractive devices, many of those considerations will be resolved by technique standardization.

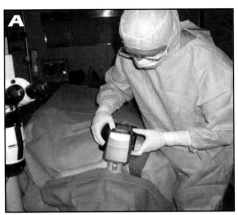 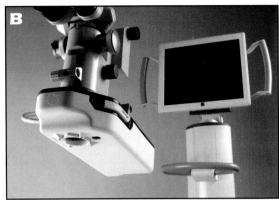

Figure 44-1. Intraoperative refractometry after cataract removal and prior to IOL implantation. (A) Portable autorefractor used during cataract surgery. (B) Microscope-integrated wavefront autorefractor (ORange).

Similar to conventional methods of IOL calculation, the intraoperative refractive technique is dependent on a number of variables that may influence the final correlation between aphakic spherical equivalent and emmetropic IOL power. While this ratio can show some variability with respect to both AL and corneal curvature, by far the most important factor is the final IOL position. A theoretical evaluation of this correlation[2] illustrates the impact of final lens position on the correlation coefficient between the aphakic spherical equivalent and final emmetropic power (ie, the theoretical ratio.) The difference between a more anterior final IOL position versus a more posterior placement can exceed 0.75 D of final emmetropia across a wide range of IOL powers. This effect is considerably larger than the one caused by variations of corneal curvature, for example.

Early Clinical Results

The first formula for intraoperative autorefraction was derived by myself and colleagues[3] in 2005 in a series of 38 eyes, 6 of which were post-prior LASIK patients. The range of the axial length was 21.4 to 25.2 mm with a range of IOL power implanted from 12.0 to 28.5 D. Autorefraction vertex distance was 13.1 and A constant of the IOL used was 118.40. A strong linear correlation was found in a series of 38 eyes across a wide range of emmetropic IOL powers (Fig. 43-2).

Using a linear regression, the following empiric formula was derived based on a strong "linear fit" between aphakic spherical equivalent and emmetropic IOL:

$$\text{Ianchulev Formula: } P = 2.01 \times ASE$$

where P = emmetropic IOL power, ASE = aphakic spherical equivalent.

In the published series, more than 93% of the variability of the final emmetropic power is accounted for by the linear relationship with aphakic spherical equivalent—in standard eyes, the conventional formulas and the optical refractive model showed equivalent predictive efficacy with a correlation coefficient of 0.96. In addition, 83% of the LASIK eyes and 100% of the normal eyes were within ±1 D of the final IOL power when aphakic autorefraction was used, compared with 67% LASIK eyes and 100% of the normal eyes using the conventional method.

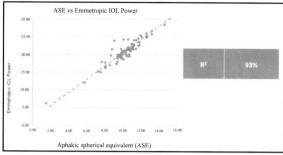

Figure 44-2. Linear regression of the original Ianchulev et al[3] series between emmetropic IOL power and aphakic spherical equivalent (ASE).

Several other studies offer additional validation of the original technique and formula described by us. In a prospective, noncomparative consecutive case series of 82 myopic eyes with a mean preoperative spherical equivalent of -12.80 D (range -3 D to -27 D), Leccisotti et al[4] derived a modified formula for high myopia population:

Leccisotti Formula: P = 1.3 x ASE + 1.45

where P = emmetropic IOL power, ASE = aphakic spherical equivalent

A more definitive study by Wong et al[5] compared the Ianchulev formula with and without a Leccisotti modification in a series of 182 eyes. This study demonstrated that while the Ianchulev formula holds across the wide spectrum of IOL powers, the Leccisotti modification performs slightly better in myopic eyes (AL >25). In addition, another set of intraoperative aphakic refractive formulas were derived from this series as follows:

For AL <25.5 mm: P = 1.97 x ASE

For AXL ≥25.5mm: P = 0.015 x ASE^2 + 1.5 x SE + 1.5

where P = emmetropic IOL power, ASE = aphakic spherical equivalent

Another more advanced development of the intraoperative methodology is the Wavetec Intraoperative Aberrometer under the commercial name ORange (WaveTec Vision, Aliso Viejo, CA). This is a Talbot-Moire based wavefront aberrometer which is attached to the surgical microscope and has been adapted for intraoperative aphakic measurements, particularly in terms of dynamic range and accuracy of intraoperative aphakic readings. This technology eliminates a significant part of the variability associated with portable autorefraction where centration and vertex can be problematic. ORange comes with pre-loaded A-constants and automatically informs the IOL power for each lens type. An additional differentiating feature is that instead of using a linear regression, ORange utilizes the refractive vergence formula in which the aphakic SE replaces the term in the standard vergence formula that incorporates the axial length. This method requires an estimation of the lens position (ELP) which ORange accomplishes with a unique formula. Preliminary results are very promising using this approach, a recent analysis of post-surgical refractive outcomes in which the most recent ORange-optimized software was used to derive IOL power recommendations, shows that in a population of 1977 non-post-refractive eyes, 81% fell within ±0.50 D and 94% fell within 1.00 D of intended target. The MAVPE for this group was 0.35 with a standard deviation of 0.48. These are compelling outcomes which are consistent with the increasing demands in the cataract-refractive space, where not ±1.00 D but <±0.5 D from emetropia is the future metric of surgical success.

Ultimately, the original methodology by Ianchulev et al may establish a purely refractive intraoperative paradigm for IOL calculation which helps solve important aspects of IOL estimation in post-Lasik eyes. It can also be applicable to the standard cataract case where refractive biometry can refine and verify the final IOL calculation. With the development of new integrated equipment that streamlines the automated refraction at the point of surgery, significantly higher accuracy can be achieved from measurement standardization, better centration of autorefraction, and incorporation of additional intraoperative parameters in the optical analysis such as keratometry. Intraoperative refractive biometry for IOL calculation may ultimately represent another important tangential point along the expanding interface between cataract and refractive surgery.

References

1. Rabbetts RB. *Bennett & Rabbetts' Clinical Visual Optics* (4th ed). Philadelphia, PA: Butterworth Heinemann Elsevier; 2007:237.
2. Sheppard A, Dunne M, Wolffsohn J, Davies L. Theoretical evaluation of the cataract extraction-refraction-implantation techniques for intraocular lens power calculation *Ophthal Physiol Opt.* 2008;28(6):568–578.
3. Ianchulev T, Salz J, Hoffer KJ, et al. Intraoperative optical intraocular lens power estimation without axial length measurements. *J Cataract Refract Surg.* 2005;31(8):1530–1536.
4. Leccisotti A. Intraocular lens calculation by intraoperative autorefraction in myopic eyes. *Graefes Arch Clin Exp Ophthalmol.* 2008;246:729-733.
5. Wong AC, Mak ST, Tse RK. Clinical evaluation of the intraoperative refraction technique for intraocular lens power calculation. *Ophthalmology.* 2010;117:711–771.

Appendix A
IOL Power Club

Members: Jaime Aramberri, MD, San Sebastian, Spain
Han Bor Fam, MD, Singapore, Malaysia
Massimo Camellin, MD, Rovigo, Italy
Claudio Carbonara, MD, Rome, Italy
Jean-Pierre Colliac, MD, Paris, France
Oliver Findl, MD, Vienna, Austria
Wolfgang Haigis, MS, PhD, Würzburg, Germany
Kenneth J. Hoffer, MD, Santa Monica, CA
Douglas Koch, MD, Houston, TX
Gabor Koranyi, MD, Vâxjô, Sweden
Scott McClatchey, CAPT, MC, USN, MD, San Diego, CA
John Moran, MD, Houston, TX
Sverker Norrby, PhD, Groningen, Holland
Thomas Olsen, MD, Århus, Denmark
Giacomo Savini, MD, Bologna, Italy
H. John Shammas, MD, Lynwood, CA

Honorary Members: Svyataslav Fyodorov, MD, Moscow, Russia (deceased)
Hermann Gernet, MD, Würzburg, Germany
John Retzlaff, Medford, OR
Rob G.L. van der Heijde, PhD, Amsterdam, Holland

Officers

2005
President: Kenneth J Hoffer, MD
Vice-President: Jaime Aramberri, MD
Secretary: Thomas Olsen, MD
Treasurer: Wolfgang Haigis, MS, PhD
EC Members: Sverker Norrby, PhD, H John Shammas, MD

2007

President: Jaime Aramberri, MD
Vice-President: Kenneth J. Hoffer, MD
Secretary: Thomas Olsen, MD
Treasurer: Wolfgang Haigis, MS, PhD
EC Members: Sverker Norrby, PhD, H. John Shammas, MD

2009

President: H. John Shammas, MD
Vice-President: Sverker Norrby, PhD
Secretary: Thomas Olsen, MD
Treasurer: Wolfgang Haigis, MS, PhD
EC Members, Past-Presidents: Kenneth J. Hoffer, MD, Jaime Aramberri, MD

Financial Disclosures

Jaime Aramberri, MD was previously a consultant for CSO Instruments, Italy.

Wolfgang Haigis, MS, PhD is a consultant for Carl Zeiss Meditec.

Kenneth J. Hoffer, MD has no financial or proprietary interest in the materials presented herein.

Tsontcho [Sean] Ianchulev, MD, MPH is the Chief Medical Officer for Transcend Medical, is a consultant for Wavetec Vision, is the inventor and patent holder for Intraoperative Autorefraction Methods, is a Clinical Assistant Professor at UCSF, and is a venture partner at Tullis Health Investors.

Susana Marcos, PhD, FOSA, FEOS has no financial or proprietary interest in the materials presented herein.

Scott K. McClatchey, CAPT, MC, USN, MD has no financial or proprietary interest in the materials presented herein.

John R. Moran, PhD, MD is a consultant for STAAR Surgical Co and AMO.

Sverker Norrby, PhD is a consultant for AMO.

Thomas Olsen, MD is a share holder of IOL Innovations (www.phacooptics.com).

Karl Ossoinig, MD is a consultant for Quantel Medical.

Thomas C. Prager, PhD, MPH is a consultant for ESI (Minneapolis, MN) and Quantel Medical (Bozeman, MT), as well as a patent holder for the Prager shell.

Giacomo Savini, MD has no financial or proprietary interest in the materials presented herein.

H. John Shammas, MD has no financial or proprietary interest in the materials presented herein.

Index

A5500, 47
A/B5500, 47
A/B P37, 46
A constant, 163
A-scan. *See* biometry, A-scan
A-scan instrumentation, 39–51
    software, 155–156
AccuSonic system, 40
Acrysof Restor IOL, 211–212
actual lens position (ALP), 150
Adjusted Refractive Index Methods, 182, 185
Advent A/B System, 40
AL-100/2000/3000, 48–49
ametropia, unilateral cataracts with, 171–172
aniseikonia, 231–232
anisometropia, 231–232
anterior aqueous depth, 27
anterior chamber depth (ACD), 119–121
    in difficult eyes, 72–73
        increase obtained by optical versus ultrasound modalities, 123–125
    IOLMaster measurement, 70
    measurement formulas, 140
    optical pachymetry measurement, 123–130
    postoperative, 149–152
    preoperative, 151
    in scleral buckle eyes, 131
aphakic eye, average sound velocity in, 19
applanation ultrasound, 34, 51, 53
    contact A-scan, 29
    versus Ossoinig immersion technique, 13–16
    velocities, 17–23
astigmatism
    correction, 213–214
    IOL tilt effect on, 223–228
    theoretical, 224
autokeratometers, 97–106
    advantages, 98
    models, 97–98, 102–106
    optics, 99
    performance, 100–101
    sources of error, 101
Aviso B, 46
axial length (AL), 9–11
    A-scan ultrasound measurement, 209
    in difficult eyes, 71–72
    effects of error in, 18
    IOLMaster measurement, 68
    laser interferometry measurement, 63–65, 75–86
    in silicone oil-filled eyes, 169
    in staphyloma eyes, 167–168
    ultrasound velocities for, 17–23
    verifying differences in, 55–56
Axis II, 46

B-scan. *See* biometry, B-scan
BESSt formula, 117, 187, 196
Binkhorst formula, 119, 138, 151, 157
Bioline, 45
biometry. *See also* applanation ultrasound; immersion technique; ultrasound biometry
    A-scan, 3
        immersion technique, 25–27, 29–36
        in staphyloma eyes, 167
    B-scan
        of staphyloma, 136
        in staphyloma eyes, 167–168
    intraoperative refractive, 241–245
    optical and ultrasound correlation, 65
    parameters, formulas for, 21–23
    in silicone oil-filled eyes, 169
    ultrasound versus optical, 151–152
biphakic eyes, axial length calculation, 22–23

CALF formulas, 23
CALF method, 21, 233
Camellin method, 182
Casio program, 157–158
cataract eyes
    A-scan, 78
    axial length measurement, 71
    biometry results with LENSTAR versus IOLMaster, 83
    high ametropic, 171–172
central corneal power/radius, 111
children
    IOL power calculations in, 215–217
    refractive growth rate in, 216–217
Cinescan B or S, 46
Clinical History Method, 180–183, 182
Colenbrander formula, 11, 119, 137, 231–232
Compact 2B, 46
CompuScan AB, 48
CompuScan LT Biometric, 48
contact lens, keratometry with, 72
Contact Lens Method, 180, 183–189
convex mirror optics, 108
cornea
    optical effective power (K), 9
    refractive index, 90–91
    scarring, IOL power calculation with, 174
    thickness
        A-scan measurement, 27
        in axial length calculations, 22
        in sound velocity measurement, 17
corneal curvature formulas, 150–151
corneal height formulas, 150–151
corneal power
    in Gaussian optics, 89–90
    keratometry indices in, 91
    manual keratometry and instrumentation in, 93–95
    measurement, 90–91
    Pentacam measurement, 115–117
corneal topography
    accuracy and precision, 112
    for IOL power calculation, 107–113, 112–113
    technologies to measure, 107–110
    uses, 107
corneal transplant, IOL power calculation with, 173–174
CSO Sirius Scheimpflug Corneal Analyzer, 189

DB-3000, 43
DB-3100, 44
DB-3000C, 43
DB-3000CG, 44
decentration measurement, 223–225
DGH 5000e, 41
DGH 5100e, 42
diopters, 90–91
double-axial length method, 151
    for scleral buckle eyes, 131–132
Double-K formula, 155, 159–160
    calculations, 205–206
    in ELP prediction, 200–202
    in ELP prediction after refractive surgery, 203–205
    ELP prediction error correction in, 205
    obtaining calculations, 205–206
Double-K method, 151, 190, 191, 199–206

Echoscan US 800/1800/2520/3300, 43
effective lens position (ELP), 9, 119–121, 139, 140
    after corneal refractive surgery, 199–204
    error correction, 205
    calculating in piggyback IOL, 220
    postoperative, 149–150, 151–152
    in scleral buckle eyes, 131
    variables in predicting, 150
emmetropia option, 171–172
Equivalent K, 187
estimated scaling factor (ESF), 139
EyeScan, 47

Feiz-Mannis formula, 190
Ferrara method, 185
Fyodorov formula, 120, 137–138, 139, 150–151, 241
    personalization, 164

Galilei dual Scheimpflug camera, 188
Galilei elevation maps, 188
Galilei instrument, 110, 111
Gaussian optics, 89–90, 116, 137
Geggel study results, 174
Gullstrand cornea, 137
    refractive powers, 89–90

Haag-Streit correction chart, 129
Haigis formula, 11, 135, 138, 140
    personalization, 164
    predicting ELP, 200–202

for preoperative ACD, 151
recommended usage, 147
Haigis-L formula, 196–197
Haigis method, 185
Hamed-Wang-Koch method, 182
Hansen Ossoinig immersion A-scan shells, 31
Helmholtz model, 94
Hi-line, 45
Hoffer formula, 119, 137–138, 151
Hoffer IOLPower Calculation Course, 4–6
Hoffer Programs®
　for computers and PDA devices, 156
　IOL power software, 131
　PDA screens, 157
　personalization screen, 165
　predicting ELP, 206
　screens, 160, 161
Hoffer Q formula, 11, 138, 139, 140, 155, 160
　accuracy, 144–147
　calculations, 161
　for computer, 157–158
　ELP errors of, 204
　IOL powers with, 172
　personalization, 164
　predicting ELP, 200–203
　recommended use, 145, 147
Hoffer/Savini LASIKIOL Power Tool, 180–181, 183, 196
Holladay formulas, 11
　formula 1, 119, 120, 138–139, 140, 151
　　accuracy, 144–147
　　for computer, 157–158
　　ELP errors of, 204
　　personalization, 164
　　predicting ELP, 200–203
　　recommended use, 145, 147
　formula 2, 121, 138–139, 155, 159
　　accuracy, 145–146
　　IOL powers with, 172
　　recommended use, 145, 147
Holladay IOL Consultant®, 158–159
　personalization screen, 165
　in piggyback IOL calculation, 220
Holladay report, 116, 117
Humphrey Topography axial map, 184
hyperopes, 171–172
hyperopic error, 220
hyperopic eyes, axial length measurement, 71

I^3 System-ABDv1/2, 42
Ianchulev formula, 243
Ianchulev Intraoperative Aphakic Refraction method, 191
immersion anterior-posterior length measurements, 57–58
immersion method, 11
　in A-scan biometry
　　avoiding pitfalls, 36
　　examination technique, 30–32
　　historic background, 29
　　instrument and probe designs, 34
　　optimal probe properties, 35
　　optimizing results, 35
　　parameters, 34–35
　　principles, 30
　　requirements, 34
　　software to improve, 33
　A-scan instrumentation in, 25–27, 39–51
　versus applanation ultrasound, 13–16
　disadvantages, 13
　fixed shell, 54–61
　versus IOLMaster measurement, 65
　with Kohn shells, 15
　with Prager shell, 15, 53–61
　　technique, 60–61
　probe, 51
　reasons for use, 29–30
　setup, 14
　of silicone oil eyes, 222
　in staphyloma eyes, 167–168
instrument error, 179
intraocular lenses (IOLs)
　after radial keratotomy, 176
　aspheric, 209–210
　axial position, 119–121
　delaying, 173–174
　different shapes, 207–208
　emmetropic, 171–172
　multifocal, 211–212
　　planning guide, 213
　position prediction, 149–150
　spherical, 207–210
　toric, 212–213
IOL power calculation
　accuracy reporting, 143–144
　adjusting, 190–192
　in aniseikonia and anisometropia, 231–232

approaches in literature, 196
automated keratometry for, 97–106
basics of formula, 3–7, 9–10
computer programs for, 156
corneal topography for, 107–113, 112
errors
    causes, 179–180
    diagnosing, 237–238
    preventing, 233–236
    treating, 238–239
formulas, 135, 149–152, 241–242
    accuracy, 144–147
    errors in, 180
    history, 137–141
    modern, 155–161
    Olsen formula, 149–152
    parameters, 163
    personalization, 163–165
    popular programs, 159–160
    predicting ELP, 200–203
    propriety, 158–159
    published in journal erratum, 158
    published results, 158
    theoretical, 149
future directions, 241–245
Haigis-L formula in, 196–197
with LENSTAR LS900 instrument, 81
for multifocal IOLs, 211–212
in pediatric eyes, 215–217
piggyback, 219–220
prediction methods, 180–189
problems for, 195
in radial keratotomy eyes, 175–176
in silicone oil-filled eyes, 169, 221–222
spherical aberration influence on, 207–210
for toric IOLs, 212–213
unpredictability of, 175–176
IOL Power Club (IPC), 5–6, 239
    members and officers, 269–270
IOL tilt
    effect on astigmatism, 223–228
    pseudophakic computer eye models, 225–226, 227
    real effect on astigmatism, 226–228
IOLMaster, 39, 50
    anterior chamber depth measurement, 70
    axial length measurement, 68
    basics of, 63–65
    cataract biometry results, 83
    description, 106
    in difficult eyes
        anterior chamber depth measurement, 72
        axial length measurement, 71–72
        keratometry mode, 72
    examination with, 67–70
    with fixed immersion shell, 54
    keratometry mode, 69
    versus LENSTAR LS900 instrument, 82–84
    measuring silicone oil eye axial length, 222
    reliability, 233
    setup screen, 73
    in silicone oil-filled eyes, 169
    software on, 156
    time of measurement, 85
iPhone/iTouch/iPad programs, 156

Jarade formula, 182
Javal formula, 90–91
Javal-Schiotz design, 94
Journal of Cataract & Refractive Surgery (JCRS), 3

K reading, equivalent, 116, 117
keratometer, 93–95
keratometric index, unreliability of, 175
keratometric power map, 110
keratometry
    automated, 97–106
    in difficult eyes, 72
    indices, 91
    IOLMaster in, 69
    manual, 93–95
    setup screen in, 73
    simulated, 115–116
keratoplasty, penetrating, 173–174
keratotomy, radial
    IOL power measurement in, 175–176
    optical zone following, 175–176
KN-1800, 49
Koch/Wang Method, 184
Kohn shell, 25–26

laser interferometry, 63–65
    axial length measurement, 75–86
Latkany methods, 190

Leccisotti formula, 244
lens-iris signal differentiation, 36
lens thickness
    A-scan measurement, 27
    formulas for predicting, 151
    in sound velocity measurement, 17
LENSTAR LS900 instrument, 11, 50
    accuracy, 81–82
    in axial length measurement, 75–86
    axial measurements, 78–79
    description, 106
    IOL calculation with, 81
    versus IOLMaster, 82–84
    measurement setup, 76
    office setups for, 77
    in optical low coherence reflectometry, 76–78
    procedure with, 79–81
    results, 84–86
    software on, 156
    technical features, 76
    time of measurement, 85

Mackool Secondary Implant method, 192
Maloney Central Topography Method, 184
Masket method, 190–191
McReynolds IOL power analyzer, 238–239
Microscan 100 A+, 47
myopes, unilateral high, 171–172
myopic error, 220
myopic eyes, axial length measurement, 71

net power, equivalent, 116

Oculus Pentacam, 186–187
Ocuscan RxP, 40
OIT-B/A/3D 1000, 45
Olsen formula, 121, 135, 138, 139, 151–152
    anterior chamber depth models, 150–151
    IOL position prediction, 149–150
    for lens thickness, 151
    for preoperative ACD, 151
operating room IOL sheet, 234, 235
optical low coherence reflectometry, 76–78
optical measurement methods, 123–125
optical pachymetry
    correction charts in, 129, 130
    examination in, 125–130
    measuring anterior chamber depth, 123–130

optical path length, 65
ORange software, 243, 244
Orbscan IIz, 108–109
Ossoinig, Karl, 10. *See also* immersion method
OTI-2000/A2000, 44

P20 A-scan, 45
PacScan 300AP, 47
Palm PDA programs, 156–157
partial coherence interferometry, 64–65
Pentacam
    calculations, 113
    corneal power measurement, 115–117
    Scheimpflug image, 109–110
phacoemulsification, 176
phakic eye
    A-scan pattern, 26–27
    calculating average sound velocity of, 17–18
piggyback IOLs, 171
primary calculation, 219–220
    secondary calculation, 220
pinhole vision testing, 237
Placido disk image, 109
Placido topography, 108
power maps, keratometric, 110, 113
Prager shell
    critical biometry tips with, 60
    with disposable tubing, 57
    how to hold, 59–60
    in immersion method, 53–61
    in immersion technique, 25–26
PreVize Optimized IOL Power Calculation Web Service, 213
programmable calculators, 156, 157
pseudophakic computer eye models, customized, 225–226, 227
pseudophakic eyes
    axial length measurement, 71
    calculating average sound velocity of, 19–20
    CALF formulas for, 23
    growth of in child, 215–216
Purkinje imaging
    of IOL tilt, 226
    of IOL tilt and decentration, 224–225

Q formula, 119

radius of curvature measurement, 90–91, 93, 110
    autokeratometric, 99–100
reflexion topography, 108
refraction error, index, 179–180
refractive effect, silicone oil, 222
refractive error, 9
    spectacle plane, 205
refractive formula, 140
refractive growth, in pediatric eyes, 216–217
refractive power, 89–91
    total, 116
refractive surgery
    anatomical changes in, 199–200
    ELP prediction after, 203–205
    intraoperative biometry, 241–245
    post-laser, 179–192
refractometry, intraoperative
    concept and methodology, 242–243
    early clinical results, 243–245
regression formulas, personalization, 163–164
retinal detachment eyes
    axial length measurement in, 131
    IOLMaster axial length measurement in, 72
Rezoom IOL, 211–212
Ridley, Sir Harold, 6
Ronje Method, 182
Rosa Method, 185

Savini-Barboni-Zanini Method, 184
Savini Method, 182
scanning slit tomography, 108–109
Scheimpflug tomography, 109–110
    of IOL tilt and decentration, 224–225
scleral buckle eyes, 131–132
scleral buckling procedure, 151
scleral shell, 25–26
Shammas No History Method, 185
silicone oil-filled eyes, 56–57
    axial length measurement, 71
    IOL power calculation, 169, 221–222
    IOL power errors in, 233
slit lamp
    optical pachymetry for, 123
    patient in, 126-128
Sonometrics DBR-100 A-scan ultrasound, 11

Speicher (Seitz) Method, 182
spherical aberration, 207–210
    longitudinal (LSA), 208–209
SRK formulas, 137–138
    accuracy, 144–147
    recommended usage, 147
    recommended use, 145
SRK/T formula, 138–139, 151, 155, 160
    accuracy, 144–147
    ELP errors of, 204
    personalization, 164
    predicting ELP, 200–203
    recommended usage, 147
    recommended use, 145
staphyloma eyes
    axial length measurement in, 167–168
    B-scan, 136
surgeon factor (SF), 138, 164

thin lens theory, 209
tilted lens
    inducing theoretical astigmatism, 224
    measurement, 224–225
    real effect on astigmatism, 226–228
triple optimization, 140

UD 1000 & 6000, 49
UltraScan, 41
ultrasonic beam
    alignment, 31–32
    parallel, 35
    perpendicular to retina, 55
ultrasound biometry, 3
    of anterior chamber depth, 123–125
    versus IOLMaster measurement, 65
    measuring axial length, 9–11
ultrasound instruments
    silicone oil effect, 221–222
    specifications, 40–50
ultrasound probes, 30
    A-scan, 25–26
ultrasound velocities
    for axial length measurement, 17–23
    in biphakic eyes, 22–23
    at body temperature, 18
    CALF method to reduce errors in, 21
    correcting error in, 21
    for various conditions, 18

videokeratoscope, 110
	accuracy and precision, 112
videokeratoscopy, 226
vitreous cavity depth, 27

Wake Forest method, 191
Wilcoxon Signed Rant Test, 85
Worst, Jan, 10

Y distance, 9

Zernike expansion, 226–227